Frank Seaver Billings

Inoculation

A preventive of swine plague

Frank Seaver Billings

Inoculation
A preventive of swine plague

ISBN/EAN: 9783337325350

Printed in Europe, USA, Canada, Australia, Japan

Cover: Foto ©berggeist007 / pixelio.de

More available books at **www.hansebooks.com**

A

PREVENTIVE OF SWINE PLAGUE,

WITH THE DEMONSTRATION THAT THE

Administration of the Agricultural Department

IS A

PUBLIC SCANDAL.

AN EXPOSURE

BY

FRANK S. BILLINGS, M. D.,

Director of the Patho-Biological Laboratory of the State University of Nebraska.

PRINTED AND PUBLISHED AT THE EXPENSE OF THE AUTHOR.

TO EVERY CITIZEN of the United States who, irrespective of party affiliations, believes in a republican administration of a government "of the people, by the people, for the people," and in an honest administration of the government, this volume is most respectfully dedicated,

By their obedient servant,

THE AUTHOR.

CONTENTS.

INTRODUCTION.

To My Readers:

The publication of this work has not been undertaken in any spirit of anger, or of self-defense, or even in advocacy of inoculation. As is mentioned in the text, every scientific investigator knows, or should know, that, so far as his investigations are concerned, no act of his but persistency can in any way hasten the acceptance of his results by the scientific world. In time they will either be confirmed or shown to be incorrect. So far as my investigations are concerned, that confirmatory verdict has been sufficiently well pronounced to satisfy any man. The recall which the people of Nebraska gave me in 1891, and their continued support, is all the necessary testimony one needs to show that where best known the public has sufficient confidence in the genuine correctness of my work and its value to it, to have it continued. Therefore, so far as the work itself is concerned, there is every ground for personal satisfaction.

But there is another side to this question, which is, the attitude that has been taken by the Agricultural Department at Washington towards the investigations carried on by the state of Nebraska, even from the day this work was first inaugurated. The question is, which side is correct, which side has been true to the public and is deserving of its confidence in this matter? One side cannot be right and the other wrong. The public must be the jury. The scientific world and the public, so far as limitedly represented by the most intelligent among the people of Nebraska, have passed the verdict in this matter in favor of the investigations of this station regarding infectious animal diseases. The National Agricultural Department not only ignores these verdicts, but does its utmost to prove them unjustified by the facts. That is its right if justified by the facts. The question is is it justified? In the following pages it will be conclusively demonstrated that the facts not only do not justify the course taken by the Agricultural Department, *but that that department has published unreliable and self-evidently false and misrepresenting testimony, and also that it has willfully suppressed testimony favorable to work done by the Nebraska Experiment Station; testimony which it is evident was and should be in possession of the Department of Agriculture.*

It is to demonstrate this fact, to show the people of the United States that their National Agricultural Department has, from the beginning, taken a course unjustified by common decency and unsanctioned by every principle of manly honor, and contrary to the constitution of the United States, that the writing and publication of this work has been undertaken by the writer, at his own expense, and not paid for out of the funds of this Experiment Station as intimated regarding other publications of the same nature in the letter to Senator Paddock of Nebraska from Mr. Rusk, Secretary of Agriculture. To that letter and the authorized publication of an infamous bulletin, known as "Farmers' Bulletin No. 8," the public must charge the infliction of this work upon it. No other course was left open to me. If our work is correct in its general direction; if it promises direct and indirect benefits to the people of the United States and even the inhabitants of the world at large, then my course is justifiable and will be so considered in time. If on the other side the course pursued by the Department of Agriculture is justified by the facts, then the publication of this work is unwarranted and will be so considered. The people must pass the verdict! Without bias, with a determined desire to present nothing but facts which are demonstrable of proof, and with that proof, I have honestly endeavored to present both sides of the question and studiously tried to show them in their full light. In order to do this, it has been necessary to introduce considerable material of an historical-personal nature, but this was called out by the personal attacks on the writer in the letters of Mr. Rusk to Senator Paddock, and to the chancellor of the University, which, *at the expense of the people*, were made public documents and sent broadcast over the land, and that to Senator Paddock in some manner found its way even to many laboratories in Europe.

This publication must not be considered in the light of self-defense on the part of its author. The condition of things do not warrant self-defense. A far greater principle is at stake. The defense of the people against the machinations and deceptions of officials honored with the highest trust. The position of the Department of Agriculture challenges the intelligence and integrity of the American people. For, if as I assert, and think I have demonstrated beyond the power of contradiction, that the National Agricultural Department boldly defies the intelligent discrimination of the people and dares to openly endeavor to "pull the wool" of deceit over their eyes in its "public documents," then the liberties of the people are threatened as they never have been before, and the letter and spirit of the constitution trampled disdainfully under foot by officials who should be honest servants of the people rather than disdainful and unprincipled spokes in a corrupt political machine. This I claim to be the exact position assumed by Mr. J. M. Rusk, secretary of the Department of Agriculture, and those who have misled him, his servants in the Bureau of Animal Industry. With the other subdivisions of the Department of Agriculture I have nothing to do, and know nothing of their work. With these words I present this Exposure of a Public Scandal in the Agricultural Department, of a government theoretically and ideally established to be "a government of the people, by the people, for the people," to the judgment of the people. Read carefully! Reflect coolly! Remember it is your liberties, your service, which is endangered. One man counts not an iota in such a multitude. It is your cause which is here presented to you, not mine. Deal fairly and justly by yourselves!

THE AUTHOR AND YOUR SERVANT.

Lincoln, Nebraska, July 4, 1892.

A PUBLIC SCANDAL.

LETTER OF THE SECRETARY OF AGRICULTURE TO HON. A. S. PADDOCK, SENATOR FROM NEBRASKA.

DEPARTMENT OF AGRICULTURE, OFFICE OF SECRETARY,
WASHINGTON, D. C., Jan. 21, 1892.

Hon. A. S. Paddock, United States Senate.

MY DEAR SENATOR: I am in receipt of extracts from three letters of citizens of Nebraska, referred by you for my information, which letters assume that Dr. Salmon, chief of the Bureau of Animal Industry, has been carrying on a controversy with Dr. Billings, of the Nebraska Experiment Station, making unjust and unwarranted attacks and interfering with the latter's scientific work. The writers of these letters request you to take such steps as may be possible to prevent such attacks in future, and to cause the chief of the Bureau of Animal Industry "to let your Experiment Station alone." Coming from intelligent gentlemen, one a member of the State Board of Agriculture, and one a state senator, these letters are deserving of attention; and as their statements are diametrically opposed to the facts, I take this opportunity to place the matter before you with some detail, since, if your correspondents had directed their attention to the State Experiment Station, *where the controversy originated*, instead of this department, the evil of which they complain might have been arrested long ago.

It may not be known to your correspondents that, when Dr. Billings first became connected with the Nebraska Experiment Station, in 1886, *a request* was made *on my predecessor that his salary be paid from the appropriation for the Bureau of Animal Industry*, so that he could have more funds to use for other purposes. It was not deemed proper to comply with that request, and since it was declined, Dr. Billings has kept up a constant succession of attacks on the Bureau of Animal Industry, not in the form of courteous, scientific criticism, but made up of dogmatic assertions, clothed in the most offensive language, and filling column after column in the *Nebraska State Journal* and *Western Resources*, but not by any means confined to these. It is a fact, which can be established by documentary evidence, that when this investigator first reached Nebraska, and before he saw a case of swine disease, he began his attacks on Dr. Salmon's investigations of this subject in an editorial covering nearly a page of the *Nebraska Farmer*, which had been temporarily placed in his charge.

These attacks were so frequent, so virulent, and so ungentlemanly in their language, and occupied so much of the space in the Station Bulletins, of which he was the author, *that it was considered a disgrace, not only to Nebraska*, but to the Experiment Stations as a whole, and resulted in a resolution of "emphatic disapproval" passed by the Association of American Agricultural Colleges and Experiment Stations. But the attacks were kept up, and he finally demanded a commission of scientific men to decide whether or not he had shown the investigation of the bureau to be unreliable, whether the reports of the bureau were true or false, and whether, to use his own words, he, himself, was or was not a fraud. On request of the National Swine Breeders' Association such a commission of scientific men was appointed by my predecessor; it was constituted in accordance with Dr. Billings' wishes; it was pronounced by him to be satisfactory; and yet, after a long and careful investigation, it endorsed the investigations of the Bureau of Animal Industry in every essential particular, and in every essential particular decided that Billings was wrong, and intimated that his methods were not all that was to be desired. Not satisfied to allow the controversy to drop here, he issued a pamphlet entitled "Evidence Showing that the Report of the Board of Inquiry Concerning Swine Diseases Was Fixed," in which he attacked the honesty and veracity of the honorable scientific gentlemen composing the commission. This disgraceful conduct, *this use of the Experiment* Station funds to carry on these bitter attacks against the Bureau of Animal Industry, must have attracted some attention in Nebraska, for it was publicly stated by the press that it was a condition of his recall that this warfare should cease.

Soon after his reappointment I received a letter from him asking that $5,000 of the funds of the Bureau of Animal Industry should be turned over to the State Experiment Station, to be expended by him in his investigations, the conditions being that the persons employed should be selected by him, and that the Bureau of Animal Industry should have no connection with the work, and no control or supervision over the expenditure of the money. At the same time he assured me that he should do his utmost against the bureau. Such a proposition, it is needless to say, was not accepted, and since that time it has not only been Dr. Salmon and the bureau that he has attacked, but myself and the Department of Agriculture as a whole.

Under the date of August 20, '91, he published an open letter addressed to me, in which he quoted from department reports, *garbling his extracts*, and misrepresenting the position of the writers *in order to deceive his readers*, and convey the impression that the department had been inconsistent in its reports. He asked that a man be sent to Nebraska to investigate his methods and the results of his inoculations; but he was careful to dictate as to the man whom he would receive, and to couple this with the threat that various influential men "are all anxious to bring this about, and I will respect their desires quietly and gentlemanly if I can, bitterly and determinedly if I must." As I declined to consider this proposition until the return of Dr. Salmon from Europe, he telegraphed me under date of September 29, in the following peremptory language, "*Parsons only will be accepted. Time expires October first.*"

A second open letter appeared soon after, which occupied four columns in *Western Resources,* and the main point of which was that he did not want a scientist sent to Nebraska to observe his methods, and would not receive one.

Still later, in the *Nebraska State Journal* of December 10, 1891, Dr. Billings published an open letter three columns in length, headed "Let Uncle Jerry Answer." "Is the secretary what he pretends to be? Dr. Billings questions the sincerity of his friendship for the farming classes," etc., etc.

Without going into a discussion of the merits of the scientific opinions involved in the controversy, you certainly will agree with me that it is gross impropriety for an investigator in an Experiment Station to attempt to dictate to a department of government what kind of an investigation it should make of a matter in which he has a personal interest, and whether it should make any investigation at all or not. And no one can fail to recognize the impropriety of an investigator paid out of an appropriation made by congress for the advancement of agricultural science, spending his time and using the resources of the station for discrediting the investigations of the National Department of Agriculture, and for abusing the secretary of agriculture, and the chief of the Bureau of Animal Industry.

In regard to the investigations asked for by Dr. Billings, I would state that I am not making investigations of the Experiment Stations, as their work is essentially separate and distinct from the scientific work of this department. If I should attempt such an investigation, however, I should not allow interested parties to dictate to me where, when, and by whom the investigation should be made.

You will find by investigating the matter that during neither his first nor second engagement has Dr. Billings been attacked by Dr. Salmon; on the contrary, Dr. Salmon has been the subject of continued attacks and abuse by Dr. Billings, and has allowed most of them to go unnoticed. When the misrepresentations and false claims of Dr. Billings have been exposed, however, and he has been driven to the wall, he finds it convenient to pose as a persecuted and long suffering individual, who has been quietly and disinterestedly laboring in the cause of the farmer. In this way, and by raising the cry of "Nebraska's interests" have been attacked, he hopes to excite sympathy and receive support. After clamoring so long for an opportunity to demonstrate the success of his inoculation as a preventive for hog cholera, it was certainly very ridiculous to find him evading the opportunity which Dr. Salmon gave him for making a comparative test at Ottawa, Ill. He wanted an investigation, but not by scientists. Here was a chance to demonstrate the value of his methods to a committee of intelligent farmers, but instead of accepting with alacrity, he did all he could to delay the test. The experiment, however, was made. One of his pupils, whom he declared to be competent, operated by his method and the result, proves, beyond a doubt, that Dr. Salmon was correct in stating that inoculation was dangerous and might spread the disease. In this case sixty per cent of the hogs inoculated by the Billings method have died from the disease communicated by inoculation, and this disease has since spread to the hogs inoculated by the bureau and also to those which were not inoculated. After his unqualified recommendation of this method and his statement that any farmer could safely use it, what more is needed to show that he cannot be relied on, and that his teachings are deceptive and dangerous.

The investigation for which he pretended to be so desirous having been made, though not under his direction or control, and having turned out as disastrously to his pretensions as the investigation by the commission of scientists made some years ago, he now turns to the people of Nebraska and frantically calls upon them through the *State Journal* to sustain him by writing to their congressmen to "stir them up," and "let them know they are your representatives and not the subjects of Jerry Rusk or his department."

The three letters which you refer to appear to be the returns from the pathetic appeal and these probably would not have been written if the witness had not been misinformed as to the origin of the controversy and as to who was responsible for keeping it up.

It will be seen from the above that your correspondents are mistaken in assuming that it is Dr. Billings who has been attacked, and that they are also mistaken in supposing that to silence Dr. Salmon would have the effect of stopping these discreditable productions from the Nebraska Experiment Station. The controversy started with and has been maintained by Dr. Billings, and he is responsible for his unscientific, undignified, and discourteous tone. *If the state of Nebraska* wishes to continue this kind of a man in such a conspicuous position, paying him $3,600 per year and allowing him to expend two-thirds of her Experiment Station funds, I suppose she has the power to do so, but her people cannot expect, and have no right to shift on outside parties, the discredit and disgrace which he brings on her fair name.

I am, very respectfully yours,

J. M. RUSK, *Secretary.*

ANSWER TO THE LETTER OF THE SECRETARY OF AGRICULTURE TO SENATOR PADDOCK.

Mr. Rusk says:

It may not be known to your correspondents that when Dr. Billings first became connected with the Nebraska Experiment Station, in 1886, a request was made on my predecessor that his salary be paid from the appropriations for the Bureau of Animal Industry, so that he could have more funds to use for other purposes. It was not deemed proper to comply with that request, and since it was declined, Dr. Billings has kept up a constant succession of attack on the Bureau of Animal Industry, not in the form of courteous, scientific criticism, but made up of dogmatic assertions, clothed in the most offensive language, and filling column after column of the *Nebraska State Journal*, but not by any means confined to these. *It is a fact, which can be established by documentary evidence, that when this investigator first reached Nebraska, and before he saw a case of swine plague, he began his attacks on Dr. Salmon's investigations of this subject in an editorial covering nearly a page of the Nebraska Farmer, which had been temporarily placed in his charge.*

So far as my attacking the work of the chief of the bureau, as published in the Reports issued by the National Agricultural Depart-partment in the papers mentioned, the accusation is true, and I have every reason as a man, an investigator, and a citizen of the United States, to glory in its truth.

The first question raised, and the one of the utmost importance in the history of the battle I have waged against dishonesty in the public service, so far as the chief of the Bureau of Animal Industry is concerned, is, *who first commenced the attack, the chief of the Bureau of Animal Industry or Dr. Billings?*

The secretary of the Department of Agriculture undoubtedly thinks it was Dr. Billings, and as perhaps many other people in the country, especially many editors of agricultural papers, think the same thing, it is now my purpose to place the entire history of my coming to Nebraska plainly before them, and as this will include rather a comprehensive history of my connection with original research, the reader must have patience. It will be shown that the secretary of agriculture has been made the victim of an unprecedented number of false and malicious misstatements, which have led him to be a most culpable deceiver of the agricultural public.

In the year 1875 I first took up the study of veterinary medicine, going to Berlin, Germany, for the purpose. When I went to Germany I neither knew what I was fitted for nor had I any direct purpose in mind, except to study. I will cheerfully admit that I had been a sort of a "vagrant," wandering over the surface of the earth "by sea and land," more with the desire of getting away from myself than anything else, being, as I had been termed, "a rolling stone," with the innate consciousness that I possessed abilities of some sort, but not knowing what they were or what they were good for. I was then thirty years old. Much to the disgust of my father's side of my family, which had occupied a most honorable position in the mercantile world of Boston, I was by nature totally unfitted to "follow in the footsteps of my fathers," even though the most satisfactory worldly success was almost guaranteed me, simply because I had neither taste nor head for business. This fact, and my unsettled roving disposition, was the cause of most profound uneasiness and unhappiness to a father whose superior never man had—a father who has been to me father, mother, and brother, all in one, and to an uncle who, having no children of his own, has been only a second father to me. There is a lesson in my life which should be profoundly studied by all parents who desire to do justice by their children. My two living relatives were merchants; their ancestors had been merchants for generations, and they could not comprehend for years, and have only lately learned the lesson, why I could not also be a merchant. Some children, in fact many, are not by any means the intellectual like of their parents in anything. Such children are to be pitied. They generally suffer under mistaken judgment and false condemnation. Parents often think they own their children to do with as they please, utterly regardless of the individual abilities of the children. I do not think my own good parents ever gave a moment's earnest thought to the matter. My father and uncle had been successful in business, and it was a natural assumption that "what was good enough for them should be equally so to me." The "boot did not fit." It pinched painfully everywhere. From youth up I had been strikingly noted for several things:

First—An innate taste for those natural phenomena which come under the head of zoological and all systems of thought relating to.

Second—A passionate love for animals, especially horses, and the more "manly" sports.

Third—An intense sympathy with misery and the (so called) "under dog" in the great social battle.

These attributes, like wine, have strengthened with age, so that now I may be called, as I have been termed, "a crank" on all such questions.

As to the first of these characteristics it was the only thing in which I excelled in school, and my teachers will bear witness that in those branches I reigned supreme without any labor on my part. As to the second and third, my record is sufficiently strong to need no further comment.

It is self-evident that a boy and man with all his abilities in those three directions was not in any way fitted for a mercantile life, therefore I first took hold of veterinary medicine simply because it was the one thing which I knew I could do, though it was much to the horror of my good relatives who would, at that time, far more gladly have supported me in do-nothingism than have the mercantile reputation of the family so disgraced as to have a "horse-doctor" a member of it, and the last male representative at that. In going to Germany, however, the "environment" was most radically changed. The scientific enthusiasm, which like a fierce fire was sweeping all before it, caught me in its devouring flames and carried me, a willing and happy victim, on with my German colleagues. The example and friendship of a Gerlach, a Virchow, a Leisering, and the men on the crest of the wave of scientific research in medicine of to-day, many of whom were my colleagues, started the fires of ambition and shaped my future course. *I determined, then and there, to devote my life to the establishment of original research in my native country and combating the miseries of human life.* With that single purpose I studied then and have since worked. Egotistical though it may seem, my nature is such that I look on every public servant, or any public man, who in any way interferes with (or endeavors to) my endeavors, as a public enemy—his personality is as nothing to me whatever. On completion of my studies as a school student, it was my secret hope that I should be able to devote my energies to the study of the infectious diseases of child-life. In that I have been disappointed, as everybody knows. In December, 1885, I was asked by the board of health of Newark, N. J., through intimate acquaintance with its then chairman, Dr. H. C. H. Herald, to take the "Newark boys," supposed to have been bitten by a "mad dog," over to Pasteur for treatment; this was a most sensational event which caused my name for the time to be most prominent in the columns of the daily press. This "adventure," to term it properly, attracted attention to me in Nebraska, though there was still another factor at work in the state, which was this:

During my early student life in Europe my friend, Professor Gerlach, then director of the Royal Veterinary College at Berlin, one day called my attention to an American boy whom he had lately admitted to the college and asked me to interest myself in him and do all I could to aid him, as he was quite young. This boy later on became a very potent factor not only in my coming to Nebraska, but in making my first term here most uncomfortable, leading to my retirement in 1889. Let me say for him, that during this Berlin epoch, and for some years afterwards, that this young man was not only one of the handsomest, most gentlemanly, and kindly of men, but apparently one of the truest I had ever known. Politics ruined his splendid qualities, as they have many another, and probably wrecked a life that naturally had many of the noblest of qualities. Myself and wife became so attached to this young man that he was as a younger brother in our family, and on our return to America became doubly dear, as he was then the only person in this country who had known and been loved by an unusually gifted child, our only son, who lies buried at the feet of our devoted friend, Gerlach, in German earth. The young man left Germany before he completed his studies and came home, graduating from the American Veterinary College, then spending six months with us in Boston, and finally settling down as a veterinary surgeon in Newark, N. J., where he also became meat and market inspector. In June, 1885, I returned from Europe, where I had been for a year studying Asiatic cholera and making myself competent in the advances which had been introduced, chiefly by Robert Koch, in the technique of investigating infectious diseases. At this time I was called to the position of pathologist to the New York Policlinic School of Medicine, which position I held until I went with the "Newark boys" to Pasteur in December of that year. While my rooms were being fitted for my laboratory in New York I spent most of the time with my friends in Newark, N. J., and one day the young man spoken of, who had become Dr. Julius Gerth, said "he would like to go west and see the country," and thought he could fix it. Much to my surprise, some weeks later, he came to me with a letter from the then commissioner of agriculture, Hon. Norman J. Colman, in which it said something like this: "That your (Dr. G.'s) name had been suggested to me as a suitable person to send west to investigate swine diseases, and that calls for that purpose had been made on the department from Nebraska, and would he (Dr. G.) write how he thought swine plague should be investigated."

As said, this letter was handed to me. I knew perfectly well that Dr. G. knew nothing of the scientific methods of investigation, as I also knew that there was scarcely a veterinarian in the country who did, and I then did a thing which I have ever since regretted and which will never be repeated, but I thought I could equalize the wrong-doing by working myself. I aided in deceiving the public, but I did it as many another has done, to aid one I loved as a younger brother. I said to Dr. G., "You go west and do your best in the field, but first come over to the laboratory and I will show you how to inoculate cultivating tubes and prepare material; then you send me both from Nebraska; make the most careful field-notes you can and I will do your scientific work and write your report for you." He answered he did not want me to do that, but I insisted on it, telling him "my own time will come later." I thought that by taking this course the public would be well served, at least as well as I could do it, and my dearest friend benefited. On condition that Dr. G. should fulfill his agreement with me, I then answered Mr. Colman's letter, giving my ideas as to how swine plague should be investigated, very nearly as I have myself carried them out since. This letter was copied by Dr. G., signed by him, and sent to Mr. Colman, who very soon sent him an appointment to go to Nebraska. That is the simple story of how Dr. Julius Gerth got to Nebraska. On arriving here he was met with the customary open-handed cordiality. The previous legislature had provided for a live stock commission and state veterinarian, and coming as he did so strongly endorsed by the Commissioner of Agriculture, Governor Dawes, who had been vainly seeking a competent veterinarian, soon offered the position to Dr. Gerth, who at once accepted it and threw up his appointment from the commissioner of agriculture. No investigations of swine diseases were made and I never received any material. Up to this time Dr. G. had scarcely made a move in his professional life without consulting with me, and so when he came east to get his family almost his first remark to me was, "I cannot get on without you and you will see that I will get you out there," or words to that effect, and no sooner did he return to Nebraska than he began to work most earnestly to bring it about, and the then regents of the University and the members of the State Board of Agriculture will bear witness that no man could possibly speak higher of another than Dr. Gerth did of me at that time. In fact, I have heard it repeatedly stated that he even went so far as to say that "*no man on earth was my equal.*" Whether true or not I have done my utmost to fill the bill. So enthusiastically did Dr. G. work that the Board of Agriculture of the state, at its annual meeting in January, 1886, passed a resolution that the regents of the State University take some steps to inaugurate the investigation of the infectious diseases of live stock, particularly swine plague. Dr. G.'s strong and energetic endorsement, coupled with the publicity given to me by the "Newark boys" case, caused the regents of the University to look to me as probably the person they desired, and the then dean of the Industrial College, Professor Charles E. Bessey, corresponded with me, and finally asked me to visit Nebraska and consult the Board of Regents at their meeting in March, 1886. In this connection the following letter is suggestive:

OFFICE OF STATE VETERINARIAN,
LINCOLN, NEB., Feb. 4, 1886.

MY DEAR DOCTOR: Whom do you think some idiots connected with the State University endeavored to get out here? Nobody but that Stalkers of Iowa, James Law of Cornell, and a fellow named Farrington from the east. You can bet I just had influence enough to sit down on that programme and did. Your appointment will probably be made about the middle of March. The Board of Regents meet on the seventeenth and would like to see you. I am just afraid that some fool will beat you out on the home stretch, but if they do it will not be my fault. Professor Bessey says that were there any funds to pay your expenses he would have you come anyway.

Yours, J. GERTH, JR.

Let me say here that I never made a single endeavor to obtain this call to Nebraska. The whole thing was the work of Dr. Gerth. I decided to visit Nebraska at the time indicated and found, as is very often the case, that about all there was to work on was a "resolution," there being but $1,200 to buy laboratory equipments and fit up rooms and for all purposes for one year. On consulting with the Board of Regents, and detailing the cost of equipment and expenses, they decided that it was beyond their power to do anything more than appropriate the $1,200, and supply one room bare of all equipment. I then made the following proposition: *I told them I wanted no position which would not promise the final completion of my ambition, the establishment of a fully equipped laboratory supported by the state; that at the end of a year they would be able to judge whether I was the man they wanted, and I as to whether the place promised what I desired to attain or not. That if they would place the $1,200 at my disposal and the room and such opportunities at the State Farm as they could give, I would put in the laboratory and pay the balance of the expenses for one year.*

This offer was accepted. Whether it was selfish or not the public can judge. This was the first and only opportunity that had been offered me, and has also been the only acceptable one, to carry out the ambition of my life. I was not seeking position, but a chance to serve humanity in the direction I had decided upon.

Mr. Rusk, the secretary of agriculture, says in his letters to Senator Paddock (it was afterwards distributed as a public document by the Department of Agriculture) that "a request was made on his predecessor that my [his] salary be paid from the appropriations of the Bureau of Animal Industry, so that I [he] could have more funds to work with." He also infers that it was on account of that request being refused that I have persistently attacked the chief of the Bureau of Animal Industry. To the last I will say that any such inference or accusation is unequivocally false.

Was there anything wrong, in the face of the fact that we were short of funds, and the commissioner of agriculture had sent a person here to do such investigations, and was constantly sending out such men, in making such a request of a national department especially organized to do just the kind of work which was about to be attempted in Nebraska? Such a proposition was made and came about in this way: It was a "condition" that faced us—a "condition" of a depleted treasury, and not a surplus as was the case then in Washington. It was myself who made the suggestion. I told the Board of Regents that I was perfectly willing to carry out the terms of my agreement, but that as I had never earned a cent to speak of in my life, it would materially aid me in obtaining private funds for our purpose if I could earn a decent salary, and did suggest that as Dr. Gerth had been sent out to Nebraska by the commissioner of agriculture, perhaps he could be induced to send me in the same way, paying only the salary, while I would still put in the laboratory and pay all the other expenses, a report of the work done also to be made to the commissioner. Was there anything wrong in that? The idea struck the minds of the regents favorably, and the president of the board at that time, Hon. Charles H. Gere, and the chancellor were authorized to take such steps in the matter as they saw fit, and I am informed, that, with the governor, Mr. Dawes, and others, did write the Nebraska delegation in congress to make such a request of the commissioner of agriculture. I can truly say, so far as I was concerned that not even a thought of the chief of the Bureau of Animal Industry came into my mind. I then left Nebraska and went to New York, packed my things and purchased additional materials which necessitated an outlay of over $800. Much to my astonishment I heard from Nebraska and also personally from the late Hon. James Laird, that the answer they received from the Department of Agriculture was that "*they would send a better man.*" In proof of the correctness of this statement the following evidence is introduced:

LINCOLN, NEB., May 24, 1892.

Dr. Frank S. Billings.

DEAR SIR: My recollection of the matter you speak of is this: At the time you came here you volunteered to supply your own laboratory and pay such expenses as could not be provided for the first year out of the small funds the regents had set apart for the investigation of animal diseases. Some of the state officials united with the Board of Regents, including myself, to ask our delegation in Washington to see if some appropriation could not be made by the Agricultural Department to your salary for that year as an investigator in this state in behalf of the Bureau of Animal Industry. Mr. James Laird, then our congressman, wrote to the chancellor that the department declined to make any appropriation and had stated that it would "*send a better man*" to do the work that we had employed you to do.

Very truly yours, C. H. GERE,
Late President of the Regents of the State University.

At Dr. Billings' request, I brought this matter to the attention of Dr. Bessey this evening, reading him the above (except the signature). He said: "That is about correct; in fact, I think it is entirely so. I *remember distinctly the last matter* about the department saying it would send a better man to do the work. That was the word that came here at that time." JAMES H. CANFIELD, *Chancellor.*
State University, May 27, 1893.

Reader, be fair! Was that an attack on me from the Bureau of Animal Industry or not? Why should its chief step at once into the arena and tell these people that he "would send a better man"? Had he such a man to send at that time? Was there one in the United States? I may be egotistical, at least I am not afraid to face the question, "is there a better man" for this work to be found, not only in this country but the world to-day? Virchow is my only master.

Had I ever done one single thing, written one word to cause the slightest suspicion on the part of the chief of the Bureau of Animal Industry that what did result would come to pass? We have some evi dence on this point. Let it be distinctly understood that I am strongly in favor of the "Bureau of Animal Industry," as a department of our government. No man in the country would so fiercely and determinedly fight any attempt to abolish it as I would. It is the nucleus of that original research for which I have given my life. No man in the country wrote so forcibly and strongly of it as I did before I knew anything critically of its work. The fault is with its work, not the bureau as a part of the government capable of immensely valuable possibilities. It is a fact that "I never saw a case of swine disease before I came to Nebraska." I was, and still am, far more interested in the human animal, and it is only as the domestic animals are valuable to man that I am interested in them. I now desire to show that no overt act or word of mine had ever been uttered or printed that justified the chief of the Bureau of Animal Industry in combating, or trying to prevent, my coming to Nebraska in 1886?

For some years previous to my coming here I had been editor of the *Journal of Comparative Medicine*, owned and published in New

York at that time by my friend Wm. A. Conklin. The pages of this journal are full of articles from me supporting what is known as "State Medicine," of which such institutions as the Bureau of Animal Industry are a part. I had never paid any critical attention to the work of the bureau, or even to infectious animal diseases, save as they bore relation to those of man up to this time. No man in the country more strongly believed in the accuracy of its reports than I then did. Like my European confreres, the suspicion that a department of any government could employ a man who would publish in its official reports absolute falsehoods for a series of years never for a moment entered my head. I did know, casually, that in earlier years Dr. Detmers had announced a "bacillus" as the cause of swine plague, and I knew that from 1880 to 1884, inclusive, the chief of the Bureau of Animal Industry had pronounced a "micrococcus" to be the cause of this disease, and I assumed this to be correct, never having compared one report with another or critically read a single page. I was quite an enthusiastic innocent in those days. I believed in the government of the United States about as a Catholic does in the Pope. My idol is shattered. It has crushed of its own innate corruption. I have strong faith in the principle yet, but a most intense contempt for the result thus far.

In the fall of 1885 the "*First Annual Report of the Bureau of Animal Industry*" came into my hands, and as editor of the journal named, and an enthusiastic advocate of State Medicine, I reviewed it in the October number of that journal, or in Vol. VI, 1885, p. 285, and here is a part of what I said, which especially bears on the point in question here:

We have before us the first annual report of this newly created bureau with reference to the contagious diseases of the domestic animals in the United States for the year 1884. It is a work which should be in the hands of every stock raiser in the country and especially every veterinarian. We recommend those of our readers who have not yet received this book to at once write the Agricultural Department at Washington for a copy. Without it they cannot know what has been done, nor what needs to be done, nor of the many difficulties which still lie in the way of doing good work. Every unbiased mind must become convinced of several things in reading this report:

First—That we, as a profession, have a most creditable representative in Dr. D. E. Salmon, chief of the bureau.

Second—That it is the best and most complete report yet issued by our government on this important subject.

Third—That a much more extensive veterinary police organization over the whole country is necessary than we have at present, and we must endeavor to make not only the United States government, but the respective state governments realize the necessity of giving employment to as many qualified men as possible if they would really do their duty by the citizens of this country.

The report in every way confirms the views we have been advocating for the past ten years, so much so in fact that many of our own words seem to have borne the spirit of true prophecy.

And in another place I say:

At last we have a veterinarian in Washington who is not afraid to publish the exact condition of things in his report.

I wonder if Secretary Rusk ever heard of the above? How I wish those words were true and were not the mistaken result of an enthusiasm born of faith in human nature that once was, but has gone never to return. What a lesson to a reviewer of such a subject—never to take one report and not consider it with its predecessors! To the honor of scientific research this reproach is only applicable to the reports of the Department of Agriculture, and, so far as I know, only those pertaining to the investigation of infectious animal diseases. Things were not running then so smoothly with the bureau as now. The press of the country was frequently opening a fusillade on it and some bad mistakes had been made, hence the above words of mine found most favorable and kindly reception from the chief of the Bureau of Animal Industry.

Let me say right here that what I am going to quote is from memory, but that I have the originals packed away where it is very inconvenient to get at them at present. Little thought I when I packed them in New York how important they might become in establishing my character in future years. My memory can be trusted, however, and what I shall quote will be very near word for word. Soon after the publication of the review quoted, the latter part of October, 1885, I received a letter from the chief of the Bureau of Animal Industry, which is in this wise:

WASHINGTON, D. C., Oct. —, 1885.

MY DEAR DOCTOR BILLINGS: I have to thank you for your review of our last report in the *Journal of Comparative Medicine*. It is very rarely that one finds one veterinarian speaking so kindly of the work of another. I have long had your exceptional qualifications in mind and am very anxious to find opportunity to make use of your valuable services in this department. Yours, etc.,

D. E. SALMON.

It will be useless for the chief of the Bureau of Animal Industry to deny that letter, for it would only put him in the same uncomfortable and disreputable position his many other contradictions have. Perhaps he would like to publish my answer. It was certainly equally cordial. I offered him about the same terms I did Nebraska, or rather the same proposition Nebraska made to the commissioner of agriculture in 1886. I did not apply for a position, as he has falsely said many times, but I did say I was anxious to serve humanity, and if it were necessary to go to hell itself or study yellow fever, that it mattered little to me, so long as I could carry out my purpose.

It is a wonder that that letter of mine has not been published. At one time it would have helped on the fight against me in Nebraska most effectually. Between October, 1885, and March, 1886, I had no further communications with the chief of the Bureau of Animal Industry, and certainly had no suspicion of any cause for ill-feeling on his part, for in December I went to Europe, and from then until I came to Nebraska, in March, I had no occasion to give a single thought to the work of the bureau or its chief. In fact I had not a suspicion that he did not have the same kindly feelings for me as expressed in his letter of October, 1885.

What, then, caused the change? What caused this man to say, or cause Mr. Colman to say, that "a better man" would be sent to Nebraska? What caused him, a public official, a man entrusted with the welfare of the live stock interests of the country, to thus attempt to deceive the people of Nebraska? He knew then, as he knows now, that that "better man" does not exist in the medical profession of this country as a pathological investigator. What was the matter, then? Fear! The cowardice of a guilty conscience! The chief of the Bureau of Animal Industry knew himself, far better than I did, in March, 1886. He knew then, as he does now, that every word printed in the Reports of the Department of Agriculture as to a micrococcus being the cause of swine plague, that every experiment reported as made with that micrococcus proving it to be *the* cause of swine plague, was false.

He knew more! The chief of the Bureau of Animal Industry is a very smart man. No man has better reason for knowing that than I have. He knows a fool when he sees one, as well as he knows how he can fool an honest yet weak man, like Jerry Rusk, by playing on his vanity and his regard for the dignity of his office. He also knows an honest man when he sees one and takes great care to keep as distant from him as possible. The chief of the Bureau of Animal Industry knew me. He knew that I was competent, and that I would find out his perfidy and expose it. That, and that alone, is the reason he would have sent a "better man" to Nebraska in 1886. This "better man" would have been a weak tool of his own, such as is every poor creature who works in the employ of the bureau to-day. Politicians have places for tools, not men. But the chief of the bureau made a serious mistake. His judgment of men goes no further than the useful or not, to him, in their character. Any appreciation of the nobler qualities of manhood is a *terra incognita* to him. He did not know Dr. Billings. When he undertook this battle he did not weigh the armament of his opponent. Life being given, the very incarnation of the Devil cannot fight a trinity composed of actual ability, incorruptible honesty, and fanatical courage in serving humanity, and these three, backed by money, will ever be too much for anything but a cowardly assassin. His Satanic majesty would have more wisdom than to undertake the job. The chief of the Bureau of Animal Industry cannot possibly comprehend that such a thing as honor in man exists. Had he not excited my suspicions by this utterly unwarranted attempt to prevent my work, to interfere with my most sacred ambition, I was (at that time, but have learned wisdom since) such a simpleton, so thoroughly good natured, that I should have written him openly and kindly that I had found these errors in his reports, and of my own findings, and asked him could we not get over them so as not to expose them? These are no idle words, written here for effect. Too many of the most responsible breeders and citizens of Nebraska have heard me frequently express them for me not to be able to confirm the statement. Too many people know that, with all my faults, a want of generosity is not one of them.

Having demonstrated that the secretary of agriculture has been wrongly informed as to the instigator of this discussion, if the publication of other side are worthy of that distinction, I wish now to show that the cause of my first attack on the chief of the bureau had no connection whatever with myself, and was not instigated by the refusal of the department to aid us in our work here. In fact, notwithstanding their proposal to send "a better man," I was so blindly unsuspicious of any virulence in their meaning that I had scarcely considered it further. But for an act of the chief of the bureau himself, it would probably have been a year or so longer before I should have examined his reports in detail and come to a knowledge of their treacherously perfidious character. Mr. Rusk says:

It is a fact which can be establised by documentary evidence, that when this investigation first reached Nebraska, and before I ever saw a case of swine plague, he began his attacks on Dr. Salmon's investigations of this subject in an editorial in the *Nebraska Farmer*.

That's clever! Mr. Rusk never wrote that. It has the hoof-prints of the chief of the bureau. It is intended to gull the public by telling them that because I had not yet seen a case of swine plague I did not know what I was writing about in criticising the chief of the bureau. It was not necessary that I should know an iota of the disease to have exposed the idiotic contradictions existing between the reports of 1884 and 1885. It required no more knowledge than to read that to-day Mr. Rusk affirms the chief of the bureau to be a black man and to-morrow, with equal positiveness, asserts him to be a white man. A similarly positive contradiction exists between the reports of 1884 and 1885.

I wonder if the secretary of agriculture ever took pains to compare the reports of his department from 1880 to 1885 with that of the latter year. If he had, he certainly could not have been politically hypnotized as he has been by the chief of the bureau. It can be said that the position of the chief of the bureau, as an investigator, is summarized in the report of the department of 1884, for he now admits, or has been forced by these criticisms, that since that time no work of investigation has been attempted by him. He says:

I may add here, that while the investigations have been directed by me, and have been carried out according to my plans, all the reports from 1885 to 1889, inclusive, so far as they relate to experiments, bacteriological observations, and post-mortem notes, have been written by Dr. Smith; and, consequently, all the criticisms of their contents aimed at me, personally, fall flat, because I am neither their author, nor have I made any change in the record of observations.—(*Journal of Comparative Medicine*, Vol. XI, 1890, p. 44.)

As all the work done by the chief of the Bureau of Animal Industry on swine diseases is now admitted by the department to have been null and void, it is evident that he has no claim whatever to any credit for original work as an investigator of infectious swine diseases. That one competent investigator has come to the same conclusion is evident from the following quotation from a critical review of the department work, Dr. P. Frosch, assistant in the laboratory of Robert Koch in Berlin, who says:

The fact stands for me confirmed that in the five years that have passed since the second plague was first announced not one single case of an independent appearance of Salmon's swine plague has been reported which can be said to be free from objections.

The mistaken conception of Salmon's place in the swine plague investigations must be laid to the fact that the publications, both on swine plague and hog cholera, have his name and as such have been quoted in the literature.

From the present publication of Smith's, however, which could not be seen in reading the reports of the Bureau of Animal Industry, it is evident that Salmon was not the discoverer of either the hog cholera germ or that of the swine plague, so now we know the true condition of things in that regard.—(*Zeitschrift für Hygiene, Koch's.*)

Perhaps this is the first time the secretary of agriculture ever had the above conclusions presented to his notice, though attention has been often enough called to them, and he seems to have been a diligent reader of the *Nebraska State Journal,* and other Nebraska papers, if one can judge from his allusion to the same in his letter to Senator Paddock.

In the report of the Department of Agriculture for 1884 the chief of the Bureau of Animal Industry told the farmers of the United States that

In former reports I have given details of experiments which, if correctly stated, *demonstrate beyond question that the microbe of swine plague is a micrococcus.* These experiments were made and the accounts of them published in advance of those of M. Pasteur's, *and the evidence furnished was all that could reasonably be required to decide a scientific question of this kind.* (P. 222.)

Of M. Pasteur's work the chief of the bureau said, in the report of the department for 1883:

M. Pasteur has recently confirmed our American investigations in a very complete manner. He shows that the disease is produced by a micrococcus; that it is non-recurrent; that the virus may be attenuated and protect from subsequent attacks, and he promises a vaccine by spring.

Now, reader, and you Mr. Rusk, if you were a plain, every day farmer and not the secretary of agriculture, what interpretation do you put on those statements? They are positive, are they not? It is that report of 1884 which I reviewed in the *Journal of Comparative Medicine* in October, 1885, in a manner so agreeable to the chief of the bureau. Certainly there was no reason for a person simply reading those reports, and knowing nothing of the disease personally, to doubt the correctness of the statements of what should be the highest and most reliable authority in the land? What possible reason could the chief of the bureau have had for doubting the correctness of his own statements? What did he fear at that time? There was not then the remotest possibility of my ever going to Nebraska, in fact I hardly knew that such a place existed then. He says:

His experiments demonstrated beyond question that the microbe of swine plague is a micrococcus, and the evidence furnished [by him] was all that could be reasonably required to decide a scientific question of this kind.

What grounds had we, then, for doubting positive assertions of that kind emanating from a department of government which, from our earliest childhood, we had been taught to look upon with veneration and to respect as the "greatest government on the face of the earth"?

I believed them to be true in the fall of 1885 and continued to think the same until some time after my arrival in Nebraska in the spring of 1886. Had I been suspiciously angry, which I was not, for I was only mildly astonished at the proposal to "send a better man" in my place to Nebraska, I had not a single known ground for attack had I desired so to do.

The secretary of agriculture, Mr. Rusk, says, and rightly, that my first attack appeared in "the *Nebraska Farmer*"; but it was not when I "first reached Nebraska," which was practically April 1,

1886. It was in the issue of June 1, 1886. The history of this "attack" is somewhat peculiar. Contrary to the secretary's assertion, as he would have seen had he read the article in question, it was not made on my own account, but on behalf of another. I have alluded with some detail, for this very purpose, to the intimacy then existing between the state veterinarian of Nebraska, Dr. Gerth, and my family and the profound attachment we then had for him. Soon after my permanent coming to Nebraska various people warned me that Dr. Gerth was not the true friend to me that I was to him, but I persistently ignored all such insinuations. I have now abundant evidence that they were true. On my return in April I found that Dr. Gerth had resigned for reasons which need not be mentioned here, and that there was quite an intense feeling against him in certain quarters. He came to me then and told me of it, and said that if I would go to Mr. Gere that I could make it all right, and that Mr. Gere could fix it all right with Governor Dawes. I did as requested. It was "fixed all right" and Dr. Gerth retained, and thereby I kept a snake in the bosom of my family which was continually sending its poisonous venom into the minds of the people. What caused Dr. Gerth to so radically change I do not know positively, yet have been told it was "inflooence" from Washington. There is no question of the intimate alliance of the chief of the Bureau of Animal Industry, the state veterinarian, and the then chancellor of the University soon afterwards, for an "offensive and defensive warfare" to "drive Billings out of Nebraska." We have so much evidence of that that it is a matter of history here and need not further be considered. I now think that it was a "deep-laid plan" to get me to write the article in defense of Dr. Gerth in the *Nebraska Farmer* of June 1, 1886, for it was well known that I was somewhat of a fighter for the right, having given Harvard College quite a taste of my mettle in exposing their fraudulent veterinary department some years previously. I now think they expected the public would turn or could be turned against me should I open my batteries, and that it would cause my dismissal. They were not so very far wrong in their hypothesis. The public did get stampeded for a while, except the solid breeders of the state, but, as is well known, it righted itself again last year, 1891, and called me back with a unanimity seldom given to a man in his lifetime. Some time previous to my coming to Nebraska Dr. A. Liautard, principal of the American Veterinary College, New York, obtained some "virus contra rouget" from Mr. Pasteur and sent some to Dr. Gerth and some to Washington, because all the infectious swine diseases were then assumed to be one and the same thing, and Pasteur had given evidence that this virus would prevent rouget by inoculation. Now it is doing Dr. Gerth no injustice to say that he was not competent to make a test investigation of that kind; but he essayed it and published something about it, though I cannot now find a copy of his publication.

In May, 1886, Dr. Gerth came to me with a copy of the *Breeders' Gazette* of May 20, 1886, in which was an article by the chief of the Bureau of Animal Industry which attacked Dr. Gerth pretty severely, and showed up the failing of his reported experiments with this Pasteur virus, as rouget was then known not to be the swine plague, having been so demonstrated the previous year by Loeffler and Schütz in Berlin. Dr. Gerth was in high dudgeon and, as I thought natural, at once came to me to defend him. As said, I now think the entire thing was put on as a cloak to draw me out. I know I told Dr. Gerth that I could not defend him, but that perhaps I could find some weakness in the chief's armor and punch a few holes in it for him.

Two essential facts have now been demonstrated which show the false information by which Secretary Rusk has been misled in forming his conclusions in regard to my attacks on the chief of the Bureau:

First—That the first attack was made by that chief on me when he asserted that he would send a better man to Nebraska.

Second—That this attack of June 1, 1886, in the *Nebraska Farmer*, was made in the defense of a supposed friend to whom I was then deeply attached.

As said above, when Dr. Gerth came to me with the article in the *Breeders' Gazette* of May 20, 1886, from the chief of the bureau, I did not know of any weakness in the reports of the chief of the Bureau of Animal Industry. I had not begun my own investigations. But it is a good rule when you can't defend your position to seek a weak spot in your enemy and at once attack that point.

Let me quote a little from the article in the *Nebraska Farmer* of June 1, 1886:

In the issue of the *Breeders' Gazette* of May 20 is an article by Dr. Salmon, entitled "Why Pasteur's Vaccine Fails to Prevent Hog

Cholera" (Let me here again quote that passage from the report of 1883). "M. Pasteur has recently confirmed our American investigations in a very complete manner. He shows that the disease is produced by a micrococcus; that it is non-recurrent; that the virus may be attenuated and protect from subsequent attacks, and he promises a vaccine by spring." Comment on this intellectual somersault on the part of the chief of the Bureau of Animal Industry is certainly unnecessary. The reader can easily see that he made a misstep somehow and fell flat on his back), in which the learned and really accomplished dilettant, who is entrusted with the responsibilities proper to the government veterinarian of so great a country as ours, enters into a polemic, *and a just one too in many respects*, against certain work that was done here in Nebraska.

The question we are going to discuss is not the correctness of Dr. Salmon's remarks, but rather, as he is so free in questioning the capacity of others for the work undertaken, to endeavor to show that the said Salmon's work leaves a wide margin of doubt open to the mind of all intelligent observers as to the amount of faith that can be had in his own experiments and publications.

In the *Gazette* article Salmon says: "Both of these gentlemen (meaning Dr. Gerth and Professor Bessey) are radically wrong in their conclusion that the French disease, called by Pasteur rouget, is identical with the American disease called hog cholera. I explained this quite fully at Chicago last November. Dr. Gerth evidently had my view in mind when he stated so positively that the germ in Pasteur's vaccine was identical with that found in hogs affected with hog cholera." In another place he says: "He (Professor Bessey) placed the material from dead hogs under the microscope, and when he says that the germs revealed were micrococci, I have confidence that his statement was correct.

Of this statement, that is, that the micrococci found in both the vaccine and the material from diseased hogs was identical, and hence the cause of hog cholera, Salmon says: "The ridiculousness of this assertion, when we come to consider the investigations that were made, would be amusing, *if it were not for the very serious consequences which follow from deceiving the public with such rash and unreliable statements.*"

I said to that in the same place:

We must certainly plead guilty to the fact that the investigations of Dr. Gerth were ridiculous failures from conception to finish; but if the assertion was ridiculous in itself, or if there is anything amusing about it, then all we have to do is to go back to the Annual Report of the Bureau of Animal Industry for 1884, and there we may read of a whole series of more ridiculous investigations, and "*experiments, which, if correctly stated, demonstrate beyond question that the microbe of swine plague is a micrococcus."* That "*these experiments were made and accurate accounts of them published in advance of those of M. Pasteur's, and the evidence furnished was all that could reasonably be required to decide a scientific question of this kind.*"

Up to that time, May, 1886, the only report of the Department of Agriculture at hand was that of 1884, from which the above remarks were taken, as the reports of the Department of Agriculture are always issued in the year later than they are dated. For instance, the report of 1891 is just out as I am writing, June 9, 1892, and that for 1885 came out about the same time in 1886. When Dr. Gerth and Prof. Bossey concluded that the micrococcus seen in Pasteur's vaccine and the material from hog-cholera diseased hogs (the government swine plague did not appear until the issuance of the report of 1886) were identical, they stood in exact conformity with the latest information then at their command from the chief of the Bureau of Animal Industry, and when Dr. Gerth came to me with the paper of Dr. Salmon in the *Breeders' Gazette* of May 20, 1886, I was of the same opinion, as I knew nothing of the address made the previous November in Chicago, and only had in mind the report reviewed by me in October, 1885, that of 1884, which, as can be seen, gave evidence that from any other investigator would be conclusive that a micrococcus was the cause of swine plague.

It was my desire to defend my friend which led me to examine the report of 1885, then just out, which he brought to me, and I cannot well explain my surprise at seeing that the chief of the Bureau of Animal Industry had completely changed his position.

The reader must now be sufficiently well aware of the fact that so far as his official publications are of any value at all, the chief of the Bureau of Animal Industry considered that he had proven a micrococcus to be the specific cause of swine plague, and this conclusion lasted until the public issuance of his report of 1885, about June 1, 1886. In this last report he entirely changes his opinion, but not in the honorable manner which should characterize every scientific investigator, more particularly one employed by a department of the government. He does not boldly say that all his previous work was mistaken or erroneous, but makes the charge in most insincere manner. He says:

Anticipating somewhat the conclusion which was arrived at later concerning the real cause of this puzzling disease, *we must say, at this point, that we no longer consider a micrococcus as the cause of all outbreaks of the disease known as swine plague.* (P. 186.)

What more evidence is necessary to demonstrate the duplicity of the chief of this bureau? Even at this time the chief of the bureau was not absolutely sure of his ground, for in the same report he says:

The bacterium which we have lately discovered, *and which we believe to be the cause of swine plague.* (P. 219.)

What was this new microbe? It was and is a bacillus.

The greater number present a center paler than the periphery. *The darker portion is not localized at the two extremities, as in the bacteria of septicæmia in rabbits, but is of uniform width around the entire circumference of the oval.*

The above description is so false that any one well acquainted with the germ of swine plague could not possibly recognize that organism in the description. Instead of "the darker portion not being localized at the two extremities, as in the bacteria of septicæmia in rabbits," the two extremities not only do color most strongly, but the whole germ can be easily colored diffusely; and it does bear a very strong resemblance to the "bacteria of septicæmia in rabbits," a fact well seen in Fig. 2, Plate X, of the special report on hog cholera issued by the department in 1889. On the same plate, Fig. 1, is given an illustration of the thing described as not coloring like the bacteria of rabbit septicæmia. I am willing to leave it to any unbiased person, especially a bacteriologist, if he would for a moment suspect that the two illustrations were intended for one and the same germ?

I have previously termed this false description and illustration a forgery, because the merest tyro cannot help coloring the germ properly. Even ordinary farmers have learned to do it in five minutes' instruction in our laboratory and have amused themselves a whole afternoon coloring these germs and examining them.

We now know that this organism was not discovered by the chief of the bureau, but by his assistant, Dr. Smith, so that it is evident that the chief has not done a single iota of original scientific work on swine diseases since his employment by the government, though he has posed before the world and willingly accepted the credit for such until his perfidy was so shown up that he was forced to acknowledge the fact. At the same time that this nondescript description was given out the presence of a micrococcus is still adhered to, for what reason none can tell, as no such micrococcus occurs as an etiological moment in swine plague. It is said:

> This micrococcus is easily distinguished by its peculiar growth on gelatine, *which it rapidly liquefies.* * * * Whether this micrococcus is a septic organism, or one which is the cause of a definite disease in pigs, cannot be answered at present. (P. 186.)

It should be known that ever since 1880 the essentials, at least of the Koch method, in bacteriological investigations have been open to everybody, and as in 1884 the chief of the bureau so positively asserted a micrococcus to be the specific cause of swine plague, he should have cultivated it and have had notes and cultures in the laboratory, and could easily have said whether this micrococcus of 1885 was or was not that of the four previous years. It cannot be claimed that this micrococcus was identical with the dipplo-coccoid bacillus which they falsely assert to be the cause of a new swine plague, for that germ never liquefies gelatine.

The date of the first demonstration of the germ of swine plague by the government investigators is impossible to be fixed, so unreliable is their testimony. They say "the two animals which infected the vaccinated pigs, as described in the preceding pages, deserve our particular attention, since they were the starting of an outbreak at the Station, *which has finally enabled us to demonstrate as the cause of the disease a specific microbe.* These two animals were brought to the Station November 4, 1885." (P. 194.) Pig 105 died on the 5th of November. Liquor cultures "*contained a motile bacterium identified later as the bacterium of swine plague.*" (P. 195.) "November 6, a number of mice were inoculated" from this pig No. 105. "Three mice received a small ulcer from the large intestine." (Ibid.) "Of three inoculated with the ulcer, one died on the tenth day; the other (No. 2) on the fifth; and the third (No. 4) on the seventh day after inoculation." (P. 196.) Note the peculiar manner in which this record is reported, the last to die being the first mentioned. Let it be understood that I am trying to fix the date of the actual demonstration that this new microbe killed a hog by inoculation with pure cultures.

Now the mouse that died from inoculation with a piece of intestine ulcer from pig 105 must be "mouse No. 1." "Mouse No. 2 died of malignant œdema." "The kidney of mouse No. 4 contained large numbers of oval bacteria *found in cover-glass preparations in the spleen of pigs later on*, and identified as the cause of swine plague. *This was the first* time these bacteria were seen since the cover-glass preparations made from the organs of pig No. 105, whence these mice had been inoculated, *proved entirely negative*." (P. 197.)

Let us fix the date up to this time: Mice inoculated November 6, No. 4 died the seventh day afterwards—November 13. *No demonstrations of the germ yet;* because "*the cultures, both solid and liquor, obtained from the heart's blood of these mice, proved, as might have been expected, to be a mixture of several kinds of bacteria.* The gelatine invariably became liquid." Such results show unusually unscientific ability, as every investigator knows. The other pig, No. 106, died on the night of 6th or 7th of November, 1885. "Cultures in liquid media and gelatine from spleen and heart's blood remained sterile, except *one colony*, in a gelatine culture from spleen, cover glasses, negative." (P. 197.) "Of four inoculations with pure cultures of the bacterium of swine plague, to be described further on, the first did not prove quite satisfactory *from the fact that the check animal died before either of the inoculated ones*." (P. 198.) That fact utterly nullifies the value of the experiment.

However, we want to get at our date for comparison with that November address in Chicago, 1885.

"Two animals, 109–113, were inoculated November 20, each with 4 ccs. of a liquid culture from the blood of pig 94," *which died November 18.* Pig died November 18, cultures inoculated November 20, pretty rapid growth and quick work for any man to have the audacity to assert "of the four inoculations *made with pure cultures*" *of any germ!* The autopsy of this pig 94 is recorded on pages 193–4, but no mention is made of any cultures having been made from its blood. Otherwise we have been told that this germ was first seen in pig 105.

No. 113, one inoculated with "pure cultures" from pig 94, died December 31, 1885. No cultures were made, as subsequent inoculation experiments had already furnished satisfactory results." Query—How could "subsequent experiments have already furnished satisfactory results"? Do words mean anything at all? Are these men imbeciles, or what are they, to insult the common sense of our people by publishing such stuff in a government report? No. 109 died January 7, 1886, when it was killed. No cultures.

The healthy check pig penned with them died December 6, after four or five days' illness. (P. 198.)

Now read :

From the foregoing description we observe that the check animal died from a very acute attack, and it seems reasonable to suppose that it caught the disease from the two inoculated animals, and being more susceptible, quickly succumbed to the virus. (P. 199.)

My patience is exhausted! Here we have traced this work of 1885 up to January 7, 1886, with absolutely no proof of any kind that this microbe of 1885 caused a case of swine plague. How then could the chief of the Bureau of Animal Industry have said anything contradictory to his report of 1884, which absolutely pronounced a micrococcus to be the cause of swine plague at Chicago, in November, 1885? How then could he have shown that Dr. Gerth and Professor Bessey were wrong in their micrococcus theory that rouget and swine plague were identical diseases, when they had his own word that they were identical in the only published report at hand? How in the name of common honorable decency and scientific honesty can the chief of the bureau claim that the germ of swine plague was discovered in November, 1885, or at any time in 1885? The fact is, these people have not yet given any scientifically exact experimental proof that the germ they discovered is the cause of swine plague, as noticed by Frosch. We did that work for them.

This exposure in this direction has gone far enough. Such political rottenness and unscientific deportment in a department of government is enough to sicken any ambitious and pure-minded man in the future of republican institutions. Only evolution keeps hope alive.

This is the stuff the secretary of agriculture, Hon. J. M. Rusk, deals out to our farmers as the truths of scientific investigation, the reliability of which he vouches for by his signature.

Let us think! What was Benedict Arnold guilty of? The secretary of agriculture is a thousand times more so. What has become of the glorious example of public honesty in high officials, set by the fathers of this republic?

The next thing Secretary Rusk objects to in the passage we are considering from his letter to Senator Paddock is, that my "attacks on the Bureau of Animal Industry" have not been couched in the language of "courteous scientific criticism." How could they be? There was absolutely nothing scientific to criticise. This question may be one of social science, of honesty in the reported work of a department of government, but it certainly is nothing else, and hence other methods and other languages must be used than that in vogue between honest and gentlemanly scientists. Further, my own work is more of a social, scientific, and political nature than that of a mere investigator. It is to establish original research in connection with state medicine in this country. I have endeavored to do honest work as an investigator, merely to show the people its value, that they might see the necessity of supporting such work. Mr. Rusk and his advisers have endeavored to nullify my last attempts. The bill for a national laboratory for the study of human and animal diseases in congress has been twice shelved by their malicious interference. As to my methods, they are my own. The people may engage me to investigate for their benefit, it is true, but I do the work in my own way in the endeavor to stimulate the people to be true to themselves. Individually, they have not yet arrived at the stage of evolution that the majority of them know enough to be this. Had they, the Bureau of Animal Industry and secretary of agriculture would not dare treat them as a pack of fools, and publish the mass of incongruities they have, under the name of 'investigations.' Were it not to do a duty which I deem obligatory to humanity, I would not be an investigator in a public position. One is too open to the cowardly attacks of unprincipled and purchasable editors, and craven politicians. It would suit me better to live in the turmoil of a great city and take an active part in the great social-political battle for justice, where I could be more free than I am even now, to scathe these political parasites and demagogic bullies now in public positions as mere tools of the bosses of the political machines. *There is but one place that a man can be a manly and independent man in this country, and that is as an editor of a large metropolitan journal, honestly devoted to the cause of justice and welfare of humanity.* Every other public office in the country is a slave's position. The political machine forges the shackles which squeeze every particle of true manhood out of the occupants. Unfortunately, the majority of such men have so little real manhood that they do not feel the galling fetters. The majority of the office holders in this country, instead of being noble men, honoring the public service, are a set of servile trucksters, political dead beats, who, for family or other reasons, have to be taken care of, because of their inability to take care of themselves. The public service of the country is a charity rather than an honor. This is the trouble in the Bureau of Animal Industry. Put in manly men and honest men as investigators, and there will be no more trouble. It is a fight for a living with such people, and not for honest investigation, which has been the spear goading them on to these malicious and unwarranted attacks on the work done in Nebraska. If my methods do not suit the people of Nebraska, I am the easiest man in the world to get rid of. They know my work has been honest and that it is reliable. They have pronounced their justifying verdict even against so mighty an egotism as that of the Hon. J. M. Rusk, secretary of agriculture of the United States of America.

Secretary Rusk further says that "these attacks were so frequent and so ungentlemanly in their language, and occupied so much space in the Bulletins of the Nebraska Experimental Station, of which he was the author, *that it was considered a disgrace not only to Nebraska, but to the Experiment Stations as a whole, and resulted in a resolution of emphatic disapproval passed by the Association of American Agricultural Colleges and Experiment Stations.*"

The people of Nebraska have answered the first part of this accusation by recalling me and again re-engaging me even since the great secretary of agriculture pronounced his verdict. Who is this man who thus boldly insults the integrity of judgment of the most intelligent breeders of a great commonwealth? Where did he come from? What education or natural qualifications has he ever displayed showing his fitness to express an opinion of work of this character? What fitness has a man such as this for such a responsible position who cannot distinguish between truth and falsehood; who will take the verdict of a hireling against the intelligent judgment of large numbers of men directly on the ground? Who is Secretary of Agriculture J. M. Rusk, or the chief of the Bureau of Animal Industry? Politically speaking the latter is evidently the most capable man. *It is a sad example of the total unsuitableness of our political machinery which puts an almost nameless incompetent into the secretaryship of a great department of government where necessity requires that such occupant shall have that degree of natural abilities and suitable technical education to allow him personally to judge of the correctness of the work of his subordinates, and not a man so unqualified as to be a mere victim of their selfish and unprincipled intrigues.*

The interests of the agriculturists of the country should be served by the most competent and generally scientific educated men in the country, with especial enthusiastic sympathies with agriculture, and not by degrading such an office by the appointment of a political heeler, because he can pull votes and mould rings.

The Hon. Mr. Rusk certainly does not know his place. He does not seem to realize that he is but a "*famulus giganticus,*" a sort of "*major domo*" entrusted by the people with watching over their general agricultural interests. The president of the United States, while honored with the most exalted and responsible position in the gift of over sixty million people, is the servant of every man of them, and hence, for the time being, the smallest man in the country. To be a faithful servant he must sink his individuality in that of the people. He is open to the criticism of every citizen. The same is true of the secretary of agriculture and all other public officials. They are worthy of respect if they are faithful to their duties in such a manly way as to command it; not otherwise. Mr. Rusk seems to think that because I also occupy a public position that I must sacrifice my right of citizenship, as most public servants do. He is mistaken in his man; that is all. Mr. Rusk has a right to criticise me if he sees just ground, as has every farmer in the country; but because I hold office I do not sacrifice my right, as a man, to criticise any other public servant, from the president down, if justice to the country demands it. Mr. Rusk seems only to be acquainted with things in office. It is time that he made himself known with the elements which make a man.

As to the opinion of the members of the Association of the American Agricultural Colleges I care nothing. Such persons are eminently respectable. I do not belong to that class. An "eminently respectable" is an individual who is absolutely negative for anything good or bad. Such are the majority of the men who expressed the opinion that I was a "disgrace" to the stations as a whole. Better be that than nothing. It may work some good eventually. "Out of evil good may come," you know, but no one ever heard of something emanating from nothing.

SECRETARY RUSK ON THE "REPORT OF THE BOARD OF INQUIRY CONCERNING SWINE DISEASES IN THE UNITED STATES."

Mr. Rusk continues:

But the attacks were kept up and he [I] finally demanded a commission of scientific men to decide whether or not he [I] had showed the investigations of the bureau to be unreliable; whether the reports of the bureau were true or false; and whether, to use his own words, he himself was or was not a fraud. Such a commission was appointed by my predecessor, and was constituted in accordance with Dr. Billings' wishes; it was pronounced by him to be satisfactory; and yet, after careful and long investigation it endorsed the investigation of the Bureau of Animal Industry in every essential particular, and in every essential particular decided that Dr. Billings was wrong. Not satisfied to allow the matter to drop here he issued a pamphlet entitled "*Evidence Showing that the Report of the Board of Inquiry Concerning Swine Diseases Was Fixed,*" in which he attacks the honesty and veracity of the honorable scientific gentlemen composing the commission.

Nothing could better illustrate the total unfitness of Mr. Rusk for his position, and the fact that the present secretary of agriculture is a pliant tool in the hands of the more clever and wily chief of the Bureau of Animal Industry than his having the audacity to thus quote and defend the report of that notoriously scandalous Swine Plague Commission. I defy Mr. Rusk to find an intelligent swine breeder in the United States who has taken any pains to observe the actions of that commission, or an intelligent and unbiased editor of a live stock or agricultural paper, who will not tell him that he also believes that that report was "fixed."

If such was and is not their opinion, how is it that I was called back to Nebraska? How is it that Mr. Rusk has not a single respectable live stock or agricultural paper in the west on his side of the question? Does Mr. Rusk think that one of the most notorious mugwumps and free-thinkers in the world, a prominent scientific socialist, and an opponent of nearly everything held sacred by the people excepting absolute honesty and cold-blooded justice, has any political pull? My strongest friends and supporters are men of his own party. I am too democratic for the democrats and too radical for the independents. Why did Nebraska call me back? Why has she re-engaged me? Was it not because her people knew that that report was "fixed," and that they had every confidence in the ability of my work? They certainly did not call me back either out of pity or with charitable benevolence. They know I am far beyond the necessity of either. There is one thing about the publication of the reports—there were two of them—of that commission which has been and is a "disgrace" to Mr. Rusk, infinitely greater than my attacks upon his pet bureau, which such papers as the *Breeders' Gazette*, the *Farmers' Review*, the *Homestead*, *Western Resources*, the *Turf, Field, and Farm*, and many others requested him to explain at the time, and yet neither he nor the chief of the bureau however deigned to satisfy that demand. The people have a right to command common honesty in the methods of their servants and that no underhand pettifogging methods be taken to deceive them or create a false impression in their minds. That Mr. Rusk consented to such a method is well known. He has been accused of it, as will be soon shown by extracts from some of the leading papers named, but has never risen to explain or deny the accusation. An honorable and creditable position for a secretary of a department in a government such as this to be in!

Two reports were issued by the commission, one by Dr. Bolton, May, 1889, the other by Messrs. Shakespeare and Burrill, dated August 1, 1889. By no political legerdemain can these reports be made to support one another. That of Dr. Bolton is actually largely, but ungenerously, in favor of Dr. Billings' position, as far as Dr. Bolton reports the results of his actual observations, which did not correspond with the reports of the bureau nor with that of the two other persons on the commission. The other report supports the bureau and condemns me, as far as it dare to, and is totally false in spirit and generally so in fact. As noted, this "majority" report was placed in the hands of Secretary Rusk August 1, 1889, *but it was sent out as the report of the commission, on slips, to the press of the country by the Department of Agriculture, and that of Dr. Bolton withheld until pub-*

lished together later. The first impression given to the public was that of the report of Burrill and Shakespeare, and, to use the expression of an editorial friend, "*Dr. Billings, you are downed.*" Soon after, a pamphlet with both reports came out from the department, and that same editor had to take "back water," much to his disgust, and Dr. Billings' stock rose, for it was evident that the reports had been "fixed." (See quotations from the *Breeders' Gazette*, soon to follow.)

Was it honorable to send out a misleading and half report when the other half had been in the hands of the secretary of agriculture for some three months previously? Was Mr. Rusk childish enough to think that honesty could be downed by any such contemptible subterfuge? The people of this country are not built that way.

Does Mr. Rusk read? Has he one person around him on whom he can depend to tell him the truth? For what purpose has he his special agricultural editor? The quotations from the agricultural press of this country and other sources have now been before the world over two years. They do not support the position of Mr. Rusk in his letter to Senator Paddock. They do support Dr. Billings. A competent secretary of agriculture should and would know all these things and not be at the mercy of a subordinate whose methods are anything but commendable. What was this commission to decide?

First—As to whether the accusation which I had made, that the announcement that a micrococcus was the cause of swine plague, in the reports of the department from 1880 to 1885 (not including '85) was false and groundless, was justified or not?

That was a very important point, but neither Commissioner Colman nor the chief of the Bureau of Animal Industry dare face that inquiry, so the question was entirely ignored in the instruction given the commission, and by them in their report; the bureau receiving wholesale endorsement in that respect by all these individuals. The instructions read:

I desire that the investigations of the board will determine the following points:

(1.) If the diseases of swine investigated by the Bureau of Animal Industry were properly described in the reports of 1885, 1886, and 1887, etc.

It may be said that in the report for 1885 it had already been announced that "a micrococcus was no longer considered to be the cause of all outbreaks of swine plague," but that was not answering my accusation. Again, it may be said that the reports previous to 1885 were not those of the bureau, but the same chief authorized them, and the report of 1884 was the first "Annual Report of the Bureau," and in it did say "that the evidence furnished (in favor of the micrococcus) was all that was necessary to decide a scientific question of this kind."

Second—The next question was, was the description of the "New Microbe of Swine Plague," as given in the report of 1885, true or false?

That question was never answered. I have shown that it was not only false, but that the germ of swine plague was not demonstrated by a particle of scientific evidence as late as January 7, 1886, and that such demonstration has never yet been given by the bureau, though it undoubtedly had the germ in its possession at that time, but was not sure of it.

Third—The next question was as to the existence of a second widespread swine plague.

Bolton never saw the germ of it in his investigations while on the board, nor in his work for months in North Carolina, but I believe he was the original discoverer of this most convenient organism.

Up to to-day we have no reliable evidence from the bureau that this organism even causes a specific local infectious disease in the hogs in this country. That it may cause complications in swine plague diseased hogs is true, not only of that germ, but others.

As I mean this answer to Mr. Rusk to be the ending of my discussion with his department of the swine plague question, and am publishing the data in detail in order to make an historical record of the facts, the reader will be interested in reading in this connection the pamphlet:

EVIDENCE SHOWING THAT THE COMMISSION WAS "FIXED."

It is evident from Mr. Rusk's letter that he should have read this pamphlet, or why should he mention it? Did he hear of it, only? It will be at once shown that had he read it, and were he the honest public servant he has been assumed to be, that he would have considered "evidence" which demonstrates that that report was "fixed." Mr. Rusk does not deny that it was not a "fixed" report in his letter to Senator Paddock. He evades an answer and tries to slip out of conscious guilt by asserting that it was an abusive answer on my part, as if one could abuse a contemptible deceiver of the people too much, when such a person is given and accepts a great public trust! Under any competent and honest government the breaker of such a trust would have had some months' close confinement for abusing the confidence placed in him. While Mr. Rusk read that pamphlet he seems to be so "innocent and bland" as not to have seen that the first person to call attention to the fact that that "report was fixed" was not the writer, but the editor of the *Nebraska State Journal*, and at the same time president of the Board of Regents of the State University, a gentleman in much better position to judge of the facts than Mr. Rusk seems to be.

The evidence furnished is as follows:

Under the heading, "The Report Was Fixed," the editor of *The Nebraska State Journal*, and who is also president of the Board of Regents of the State University, published the following in a late issue:

"The suggestion made by *The Journal* after the report of the swine disease commission was published last summer, that some undue influence was exercised on the members thereof to induce them to suppress all that they had really found out by their protracted investigation, is pretty well borne out by a letter addressed by a member of the commission to Dr. Detmers, of Columbus, Ohio, on the day before the report was published. The letter was as follows:

"'AGRICULTURAL EXPERIMENT STATION,
"'UNIVERSITY OF ILLINOIS,
"'CHAMPAIGN, ILL., Aug. 13, 1889.

"'MY DEAR DR. DETMERS: I suppose you have seen by the papers that the report of the "Board of Inquiry Concerning Diseases of Swine" has been presented to the Washington authorities, and I suppose too that you are disappointed at least in my signing the clause in respect to your early work. Yet I know fully that you would not feel hard if you knew what I did do and what I tried to do and failed. I could not help knowing that the germ you worked with 1878–80 was the same as that now held by you to be the cause of the disease you call swine plague. I knew this without examination this last winter. I have not said you are not entitled to priority in this, neither has the board. After a little, when the time seems to be proper, I want to say as an individual what I know and what should have been said in this report.

"'While I could not help doing what was done, I was far from satisfied.

"'As to there being two diseases with different germs we were all agreed, but not satisfied as to the commonness of occurrence of the one called swine plague by the Washington folks. We found one outbreak away from the east where it had only been found near Washington and Baltimore. It is not probable that it is a prevalent and serious malady, though it seems bad enough when once introduced to a herd. The most that troubled me about it was whether or not it was a peculiar variety of the common germ. But it has so many points of difference and holds them so tenaciously in all process of culture that it seemed to me impossible to consider them as one thing in any sense. This one new outbreak was in Kentucky. I wanted to go back and further study the disease there, but neither had opportunity or permission. Hoping to see you soon, I am faithfully yours,

"'T. J. BURRILL.'

"The language of the letter is positive proof that the commission was making a certain sort of report under compulsion, and as the commission was acting under the authority of congress and was paid by an appropriation made by congress, *The Journal* calls for a congressional investigation of that report which has cost the country such a smart sum.

"Let the committees on agriculture in the senate and the house at once look into the matter and see what screw was loose and see who had sufficient hold on the members of the commission to force a crooked report in favor of Mr. Salmon's alleged 'germs.' Also how it came that the report of the commission was to the effect that there were two 'widespread' infectious swine diseases, with two different germs, when as a matter of fact the commissioner owns up that there was no proof submitted that the so-called 'swine plague' prevailed to any extent anywhere and was not found anywhere except 'near Washington and Baltimore.' Let congress clear up this crookedness at once, for no more important matter can engage the attention of that body just now than that of the diseases of our domestic animals."

As the editor of *The Journal* notices, this letter was written by Professor T. J. Burrill on August 13, 1889. Attention has been frequently called to the fact that advance sheets of a part of this report only, the so-called "majority report," were sent to certain of the agricultural papers of the country a week or so in advance of the distribution of the complete document; the public being first informed of it in the issue of the *Breeders' Gazette* of August 14, 1889, which paper, as well as others, was so misled by this advance flyer, that it had to recede from the tenor of its editorial in its issue of August 14, and in that of August 21 asked some very pertinent questions which have not yet received their answers.

An interesting question is, why should Professor Burrill feel it necessary to write such a vacillating and self-condemnatory letter to Dr. Detmers, even before any part of the report of the commission, of which he was a member, had been given to the public? Is it not positive evidence that Professor Burrill felt the weakness of his position in that report? Is it not certain that he felt himself guilty of doing a great injustice to Dr. Detmers, as well as to the hog interests of the country? Does it not show that he really wanted to be an honest man, that really he is one, but that lacking backbone he was forced into doing things directly contrary to his own knowledge of what was true and right?

Just at this point I desire to publish a letter which especially bears upon this point. It is also directly related to the previous one by Professor Burrill. I telegraphed Professor Burrill August 10, 1889, to know what had become of the Nebraska hogs which the commission took east, and received the following reply:

"CHAMPAIGN, ILL., Aug. 10, 1889.

"DEAR DOCTOR: Have just answered your telegram to the best of my ability. When I saw the pigs no trial had been made upon the Nebraskans with Salmon's 'swine plague,' and I am quite sure nothing was said about it to me by Dr. Shakespeare subsequently. *Abundant trial had been made both by feeding and by subcutaneous inoculations with the Washington 'hog cholera without serious effect. None stood the test so well as those Nebraska pigs.'* It is, however, quite possible that Dr. Shakespeare has tried 'swine plague' (Salmon's) since our correspondence upon the subject. Have not seen anything in print or report. Am not certain what publicity has been given. The two diseases are acknowledged, the one, however, said to be much more prevalent than the other. No attempt is made in the report to compare European work, though we did have some cultures from abroad.

"You can understand that it was difficult for separate workers to make a report. Would like to have published full details, but found only certain conclusions could be agreed upon, not always worded as one would do it for himself. At any rate, nothing was considered except what is the fact.

"Very truly yours, T. J. BURRILL."

From these two letters it is very evident that there was no uniformity of opinion among the members of that "commission," as also from the minority report of Dr. Bolton, who did not say much in favor of a second "widespread epidemic of hogs in this country," as both Burrill and Shakespeare did in their majority report, where they falsely say: "It is the opinion of the commission, based upon their own individual observations and examinations of the subject, that there are at least two widespread epidemic diseases of hogs in this country." The point of importance to the hog raisers hinges upon the words, "two widespread epidemic diseases among hogs in this country," one of which has been "called by the bureau authorities 'swine plague.'" Attention has been frequently called to the fact that the commission did not examine a single hog in Illinois, Indiana, Iowa, Kansas, or Missouri, and I have asserted time and again that they only examined properly one single hog from a natural outbreak of hog disease in Nebraska, and I have lately been told that they did not make a single examination in Ohio, but simply looked over some of Dr. Detmers' slides and preparations. But to add insult to the injury this commission has endeavored to do the hog owners of the country, we now find, by Professor Burrill's admission in his letter to Dr. Detmers, that this second "widespread epidemic disease among hogs in this country has only been found near Washington and Baltimore," and by the commission "in Kentucky"; but in how many hogs "in Kentucky" we are not informed. We do not know whether it was a herd outbreak or only one or two individuals; in fact, we know nothing about it, and they do not think it of sufficient importance to mention it in their report. In fact, not having mentioned this disease in their report, and not having a particle of experimental evidence from the commission in favor of it, we can say that neither as individuals or as a board of inquiry have they given a particle of evidence that they ever saw a case of that disease. Assertions without evidence do not count in science.

On the other hand, we have other evidence that this second "widespread epidemic disease" is not an epidemic at all, and nothing but a very occasional complication or disease, which but seldom occurs anywhere in the country. In Bulletin No. 6, new series II, July, 1889, on Hog Cholera," of the South Carolina Experiment Station, Dr. Bolton says:

"The following investigations were begun in November last year, and have been continued up to the present time (about seven months). In the course of our work we have visited many widely separated portions of the state and encountered the disease in all its stages."

We would call especial attention to the last assertion, and hence repeat it: "In the course of our work we visited many widely separated portions of the state, and encountered the disease in all its stages." This report then says: "In our investigations we have tried to discover the kind of germ or bacteria which caused the disease. We visited various portions of the state and took along all the best apparatus and appliances for our work. Material collected in this way was brought back to the laboratory and thoroughly examined. The germs which we found were cultivated and grown on suitable materials and were tested upon mice, rabbits, and hogs. We obtained them from the liver, spleen, blood, and intestinal organs. In some cases we failed to get any germs at all, and although we obtained, in other cases, several different kinds of germs, we have no reason to believe that more than one of them is concerned in the disease."

It would then seem that examinations of such a character, which extended over a period of about seven months, and to outbreaks in so many parts of the country, added to those made by the author (Dr. Bolton), as a member of the commission, should have been extensive enough to find one single case of the second "widespread epidemic disease," which being "swine plague" should from its name be a pest of great danger to the farmers of this country, but Dr. Bolton's examinations lead, happily, to quite a different result, for in this S. C. report he says, after describing the organisms he found, that "the only organism deserving special attention is one which we regard as identical with the bacillus of hog cholera of Salmon," and in his report, as a member of the commission, he says:

"During my work as commissioner I have failed to meet with an epizootic which I am satisfied was what is termed swine plague, in the bureau reports, though previous to my appointment on the board I studied one such outbreak. In this case, however, I directed my attention to the bacteriological questions exclusively, and I am therefore unable to pronounce on the difference in the pathological lesions in the two diseases. But I am not inclined to attach any great importance to these differences as set forth in the reports. The descriptions otherwise I find correct and well stated. In my investigations as commissioner I have been able to find but one organism, which, in my opinion, caused the outbreaks under examination, and that I regard as identical with the hog-cholera germ described in the reports of the bureau, and I find the description therein given correct. As will be inferred from what has gone before, I feel sure that another organism, correctly described in the reports as the 'swine-plague germ,' is found under circumstances which render it highly probable, if not certain, that it also causes disease.

"As to whether these two organisms are always present and operate together to cause disease, or whether the two are merely varieties of the same germ, must be decided by future investigation. The differences between them, as pointed out by the bureau, are sufficient to compel us to treat them as different germs, however perplexing it may seem that two micro-organisms are capable of producing such similar or, it may be, identical lesions."

As to the other organism (Salmon's swine plague), we wish to call attention to some very striking inconsistencies in Dr. Bolton's remarks:

First—He thinks it also "causes disease." Suppose it does. There are several other germs which do cause lesions in swine plague (hog cholera), and one which causes very marked lesions, but yet which have no necessary connection with the plague, which they did not discover,

4

though it was shown to them in Nebraska. There is a vast difference between causing a local, non-extending disease and a pest or plague. But Dr. Bolton is not sure even of this, for he says further on that he does not know "whether the two are merely varieties of the same germ" or not.

Again, in one place he says, "but I am not inclined to attach any great importance to the differences (in the pathological lesions) as set forth in the reports" of the Bureau of Animal Industry, while in another he absolutely contradicts himself when he says "the differences between them (hog cholera and Salmon's swine plague) as pointed out by the bureau are sufficient to compel us to test them as different germs, however perplexing it may seem that two germs are capable of producing such similar or, it may be, identical lesions."

Now this may at first appear as unintelligible as Choctaw to some of our readers, but a little close reading will show that two members of this commission were having a very severe struggle with their consciences while making their report. In a late publication on swine diseases by Professor Welch, of Johns Hopkins, he asserts that not a single undoubted epizootic outbreak of this second swine plague has yet been seen in this country.

And Salmon said, in his address before the National Swine Breeders' Association, "then there is the germ in hog cholera, a disease widely distributed, generally extremely fatal, *and probably productive of the greater part of the loss which falls upon the hog raisers.*"

After such an admission, is it not self-evident that the government is misrepresenting when it talks about a "widespread epidemic disease" which it calls "swine plague?"

To return to Professor Burrill's letter to Dr. Detmers, upon a point which would have neither interest nor value to the farmers, did it not show most conclusively that Professor Burrill was terribly conscious that he had given his support to a false statement. What matters it to the farmers who first discovered the germ of hog cholera? Certainly nothing. But if a man can give false evidence in one direction, it is to be assumed that his veracity cannot be trusted in any other.

Why should Professor Burrill have been so anxious to apologize to Dr. Detmers for his actions, even before the report had been given to the public?

It may not be known to but few of our readers that Professor Burrill worked more or less with Dr. Detmers when the latter was studying swine plague at Champaign, Ill., and this is why Burrill wrote Detmers: "I cannot help knowing that the germ you worked with 1878-80 was the same as that now held by you to be the cause of the disease you call swine plague. I knew this without examination this last winter. I have not said that you are not entitled to priority in this, neither has the board."

Let us see what they did say:

"It is the opinion of the commission that the microbe that Dr. Detmers at present regards as the specific cause of 'hog cholera' is probably the same microbe which is considered by the bureau authorities as the specific cause of hog cholera; but, according to the present requirements of bacterial research and interpretation, it is impossible to declare that the organism as described by him in his reports published by the Department of Agriculture was the same thing."

Dr. Bolton says about the same thing. But these men knew that they were giving rise to a false impression in order to support the tottering fabric of the Bureau of Animal Industry, which they did as follows:

"It is the opinion of the commission, based upon their own individual observations and examinations of the subject, that there are at least two widespread epidemic diseases of hogs in this country, which are caused by different micro-organisms, but which have clinical history and pathological lesions more or less similar, and very difficult to distinguish without the aid of a microscope and resort to bacteriological methods; and that these two epidemic diseases have been fairly well described in the recent annual reports of the Bureau of Animal Industry."

In all honesty a man's work must be judged by the light of his time, and if, as Burrill admits, Dr. Detmers did have the same germ in 1878–80, then the latter should have full credit for it, and all honest investigators have cheerfully given such credit where it was deserved, as in the case of Brauel and Pollender, who discovered the germ of Anthrax in 1855 and 1857, respectively, and is the case to-day as regards the noted Robert Koch and his germ of human cholera. But Salmon seems to have been fully satisfied with the verdict of the commission in the case, for he says, in a late publication:

"To my mind it is not the discovery of the germ in the liquids of affected animals which entitles a man to credit, but it is the isolation of the pathogenic species, its accurate description and the production of experimental evidence to show that it causes the disease."

The bureau has not yet given any such exact evidence in support of its position in swine plague or any other disease that it has investigated.

This is all very true; but supposing the discovery and description are such that any honest man can and must recognize their correctness, what then, when modern methods of isolation and experimentation had not been developed?

Again we ask, had Brauel and Pollender done any of these things with bacillus anthracis? Has Robert Koch completed these conditions beyond question with his celebrated "komma"? Has Dr. Salmon given one single description of the germ of hog cholera which did not "differentiate" from some other given by the same author? But we forgot. Dr. Salmon now admits that he has done absolutely nothing, and though the "commission" almost strangled themselves to give him credit, he now washes his hands of all credit and blame and puts it on an unfortunate scarcely mentioned in any of the reports up to that time, and absolutely not in connection with the discovery of any germ.

Dr. Salmon does it in this manly and honorable manner in the *Journal of Comparative Medicine*, January 1, 1890:

"I may add here, that while the investigations have been directed by me, and have been carried out according to my plans, all the reports from 1885 to 1889, inclusive, so far as they relate to experiments, bacteriological observations and post-mortem examinations, have been written by Dr. Smith, and, consequently, all the criticisms of their contents aimed at me, personally, fall flat, because I am neither their author nor have I made any change in the record of observation."

Why, then, did this same Dr. Salmon say in the same journal April, 1888:

"I have sent to the leading investigators of Europe cultures of the germ of hog cholera, together with copies of my report," and again in the same place: "Early in 1886 I published a series of articles, in which I demonstrated that hog cholera was a distinct disease and a very different disease from rouget or rothlauf," which fact he only copied from the work of Schütz and Loeffler in Germany, published many months previously? Why then is the "I" of the Bureau of Animal Industry so suddenly attended with this wonderful degree of modesty and self-abnegation?

It would be interesting to the public to know what manner of man this Dr. Smith can be to thus be made a tool and foot-ball of by the chief of the Bureau of Animal Industry.

That is the whole of that document. One wonders if Secretary Rusk really ever read a word of it? What can he say to Burrill's letter to Dr. Detmers? Is there not sufficient evidence in that letter, that something was 'fixed,' to condemn any man pretending to be a scientist to everlasting condemnation? Must the country take this as a fair and honest sample of the kind of men Mr. Rusk supports in office? Mr. Rusk in his letter tells us that this is the kind he believes in. He tells us that I attacked "the honesty and veracity of the honorable scientific gentlemen composing the commission." Mr. Rusk's judgment of what is honorable and true must be somewhat biased.

"*After a little, when the time seems to be proper, I want to say as an individual what I know and what should have been said in this report.*" Professor Burrill said that of the May report to which he put his name. He also said: "*While I could not help what was done, I was far from satisfied.*"

"After a little" is now almost three years, and still the world waits for Burrill to tell "what I know" which would be especially apt at this time. It will never be told. Burrill dare not tell the truth. He could not stay in his position did he do so. Does any one doubt that the members of that board were 'fixed.' Does any one think that Burrill and Shakespeare ever made that report as printed? It is not a scientific report. It is a mass of assertions unsupported by an iota of evidence in any single direction. It is a political lie from end to end. The government cannot support it. Secretary Rusk endorses it.

Who is Secretary Rusk as an authority? Let us see what those who have read the report, and are competent to express an opinion, have said.

REVIEWS

OF THE

Leading Live Stock Papers of the West

AND

Eminent Scientific Publications

ON THE

REPORTS OF THE "BOARD OF INQUIRY."

EVIDENCE THAT RESPONSIBLE AUTHORITIES AND EDITORS OF AGRICULTURAL JOURNALS ENDORSE DR. BILLINGS' OPINION.

To any one acquainted with the inside history of this controversy between Dr. Billings and the Agricultural Department at Washington, it is evident that the secretary of agriculture has been misled or is entirely ignorant of the opinions that have been expressed as to the report of that now notorious Board of Inquiry. That that "report" was never accepted, either by a single medical authority who has taken pains to read the whole literature or investigate the subject, or by a single honorable live stock or agricultural journal, is a fact so notorious that the silence in its favor is appalling. In order to enlighten the secretary of agriculture and the public, the following evidence, which is believed to be all there is or has been published, is herewith given:

The Agricultural Department of the United States named a commission, composed of Shakespeare, Burrill, and Bolton, to settle the question between Salmon and Billings, in connection with the pestiferous diseases of swine in the country, and to make proper investigation, and report on the same. The commission could not complete the task in the time given them, which fact they mention in their report. Nevertheless the government published the report, and accepted its conclusions. The individual experiences of the commission in the diseases in question are small, especially insufficient, and of extremely problematical value are they in relation to the "swine plague" (Salmon's) or infectious pneumonia. Notwithstanding the assertion of the commissioners that they based their conclusions on personal observations, and notwithstanding certain positive statements in their "conclusions," the report gives the impression of uncertainty for want of certainty and a sufficient scientific foundation. The conclusions appear, therefore, to be statements which have only the worth of insufficiently founded suppositions, as their correctness cannot be substantiated except with difficulty. * * * As to the priority of the discovery of the germ, and the quality of the scientific work, the commission took the side of Salmon as against Billings, but also took care to mention that the government could go on with the work without any outside help. Billings' preventive inocculation finds acknowledgment between the lines.—(*Jahresbericht über die Fortschritte in der Lehre-Pathogenen Mikro-organismen*, Baumgarten, 1889—Annual Report on the Progress in the Study of Pathogenic Germs, 1889, p. 178.)

In remarks on some of Billings' publications while in Chicago, the succeeding *Jahresbericht*, 1890, says:

These and other things form the introduction of a pamphlet (by Billings) against the assertion of the swine plague commission that his methods were not correct. He justifies the methods that he used, the spleen of swine, to gain cultures from, and demonstrates that it was carefully done, so that it appears as if no possible objections could be raised against it. Billings blames, apparently with justice, the commission for treating the subject in a very superficial manner, so far as his laboratory and his work was concerned. (P. 185.)

In a very careful and critical review not only of the American swine plague literature, but also a detailed comparative study of the germs of the true American swine plague (not Salmon's) with those of swine plague (European) and other diseases of Europe, Dr. P. Frosch, assistant to Robert Koch in the Laboratory of Hygiene, Berlin, Germany, published his conclusions as follows:

1. The bacterium of hog cholera (Salmon) and swine plague (Billings) are identical.
2. The same is the cause of the American swine plague, while the proof of the etiological (causal) connection to this pest, of Salmon's germ of swine plague, that is for the existence of a second ("widespread epizootic disease") plague of equal extent is not sufficient to warrant such a conclusion.—(*Archiv für Hygiene*, Vol. 9, p. 279.)

On page 247 of same journal, and in the same article, Frosch says:

All these circumstances (previously considered in detail), that is the appearance of cholera and swine plague (Salmon) in the same animal and the same outbreak, and the same time, the presence of still other germs in the organ of the deceased animals, and the fact that in such cases reference is made to chronically diseased hogs, as well as the results of the inoculation experiments, justify the assumption that the swine plague bacterium of Salmon is an accidental presence in chronic cases of hog cholera, for which any idio-pathogenic relation to another pest not sufficient evidence has yet been produced.

As the members of the Swine Plague Commission saw fit to condemn the methods by which they supposed I worked all the time, and did say I worked with at the time, as they brought the best practical-results, it may be well to state what an entire outsider has to say on the methods of investigation as published by the Bureau of Animal Industry in its own reports. On this point Frosch says:

The most important results must be those obtained in swine by the inoculation of undoubtedly pure cultures of the hog-cholera (S.) bacterium, because in that way only can the specificity of the organism be proven. *But it is directly in this relation that one meets with essential insufficiencies (in the government work).* Notwithstanding the numerous experiments made by Salmon with pieces of organs and heart's-blood, etc., of diseased swine, still those made with pure cultures of the germ are made prominent by their scarcity and generally negative character. The small number of these positive results would not call forth any objections as to their value were it not that they do not fulfill those conditions which are essential in deciding questions of this kind. Of the experiments detailed in reports of 1885 and 1886, *there is not one which can be said to be free from objections.* The cause of the same lies, first, in the fact that the order of the experiments was not correctly arranged, the control animals dying first in two cases and at the same time in a third; while in the feeding experiments of the same year control animals are not mentioned. A second very essential objection to the method pursued by Salmon is the manner he went to work to obtain the pure cultures, * * * the consequences of such an unscientific method are to be distinctly seen in the report itself, when mixed with the inoculated germs, other "large" or "fine" bacilli are mentioned as being present. The best proof of the unscientific character of his method is given by Salmon himself in the assertion that cholera and swine plague can occur in the same animal, a connection which absolutely demands in each case the use of the plate method. (*Ibid*, p. 243.)

To this detailed study and exact report of Dr. Frosch's the government investigators most seriously objected, as would be natural, and one of them made reply thereto in the same Archiv (*Zeitschrift für Hygiene*, Vol. 10.) To the same, Frosch answered as follows:

The foregoing article by Dr. Theobald Smith, and especially his comments on a contribution of my own, induces me to once again place my position regarding the American swine plague clearly before the world. To begin at once with the point that seems to have mostly irritated Smith, I do not think that any one but he can find in my former article any special partiality or unjust discrimination in favor of the investigations of F. S. Billings. Such an estimate of my contribution to the question is only possible to a person who has cursorily read the same, or who is profoundly ignorant of the character of the hygienic institute at Berlin, and the work which is done therein.

As I declared in the beginning of my previous article, the reception of the cultures from Billings at this institute, and the request of Prof. Koch, led me to enter upon the study of the American swine plague. The task which I had to undertake was not, as Smith appears to believe, to decide as to whom belonged the most or earliest credit for work done, but to see how far, from a purely scientific point of view, the solution of the real question of the etiology of this disease had been advanced, which, at the time, seemed to be buried in darkness, in consequence of contradictory publications.

That I should depend more upon my own investigations with the cultures at my disposal than upon the investigations of Billings should not be questioned by Smith. That I should refer to the investigations of Billings, after assuring myself of the identity of his germ and Salmon's hog cholera germ, in order to decide as to the pathogenity (disease-producing power) of the germs in swine, was forced by Smith himself, *for, much as I regret to have to repeat it, the methods and experiments published in the reports of the Bureau of Animal Industry (1885 to 1887–8) do not correspond to the scientific conditions necessary to the establishment of a new infectious disease in a manner to be desired.*

It is by no means necessary for me to repeat what I said in my previous publication, as Smith admits it, in regard to the report of 1885, and, on the other hand, the superficial investigations described in the other reports display so little exclusion and exact employment of Koch's methods to correspond with the importance of the assertion of the appearance of a new exciter of infection (germ) closely related to the swine plague.

I might here call attention to the fact that even in the work of Smith, described in his article in this issue of the *Zeitschrift*, the method for differentiating the two germs, the employment of the hanging drop, is not sufficiently reliable.

Smith also seems to complain that I have considered the mentioned reports of the Bureau of Animal Industry too closely after the stand point of to-day. Even though we may have to-day new ideas of what a pure culture should be, or substantially other methods of obtaining the same than formerly, still it was perfectly justifiable to prove the case as to how the earlier results of the bureau correspond with these newer ideas.

As shown in my previous publication, it is evident from the report of the bureau that at that time Koch's methods were well known there, and I do not think that it is demanding too much of the bacteriologists of the bureau to assume that they are acquainted with the methods as published in their reports.

As to the slur upon Billings' work in Smith's publication, I can safely leave it to the former investigator to consider them. I have only to refer here to his inoculation experiments in swine, to which I am inclined to give full credit as reliable evidence, because Billings emphasizes the control of the pure cultures used in the same by other cultivating tests.

The number of these experiments was sufficiently great to demonstrate the infectiousness of the germs and their specific characteristics. Smith neglects to observe that the experiments quoted by Billings were specially selected out of a great number. I find it remarkable that Smith should question these twelve published experiments of Billings, the value of which is beyond doubt.

The inference (by Smith) that I allowed Billings' publication to influence me in the consideration of Salmon's swine plague by no means corresponds with the facts. For judging this question I have referred to the publications of the bureau on the assumption that in these official reports the most reliable material must be found.

In regard to the question as to whether the Salmon swine plague germ is an independent cause of a specific disease, I cannot change my previously expressed opinion. How far I am right can best be judged by reading Smith's publication in this journal.

As will be remembered, Salmon, supporting himself on the German Schweine-seuche, distinctly and emphatically asserted the existence of an independent plague and at the same time united with the hog cholera, which had the same degree of extension over the country.

Judged by the investigations published in the reports of the bureau, we find but proportionately few cases, and these not free from objections, of the appearance of a disease-producing germ in chronic cases of hog cholera. The conditions closely resemble those seen in certain infectious diseases of man, where the secondary appearance of pathogenic germs has long been observed, without any one asserting them to be independent causal moments, or the cause of extensive epidemics.

The last two cases reported by Smith in favor of his swine plague do not give evidence which sufficiently excludes the concomital existance of hog cholera also. At the same time and the same place, several swine are reported to have died from the cholera.

The fact stands for me confirmed that in the five years that have passed since the second plague was first announced, not one single case of an independent appearance of Salmon's swine plague has been reported which can be said to be free from objections.

The mistaken conception of Salmon's place in the swine plague investigations must be laid to the fact that the publications, both on swine plague and hog cholera, have his name, and as such have been quoted in the literature.

From the present publication of Smith's, however, which could not be seen in reading the reports of the Bureau of Animal Industry, it is evident that Salmon was not the discoverer of either the hog cholera germ or that of the swine plague, so now we know the true condition of things in that regard.

An American reviewer of the entire question, so far as it pertains to the publication of the Agricultural Department, and the report of the Board of Inquiry, says:

We have compared the extracts cited by Dr. Billings from the reports of the Department of Agriculture, line by line, and word for word; we have carefully examined the plates accompanying these reports to which he refers, and we have also read the entire context from which he quotes, in order to avoid any possible bias, which may follow from reading disconnected sentences. As a result we are enabled to say that all of Dr. Billings' charges, sweeping and severe as they are, are true and just, and we cannot understand how the committee appointed to investigate the special question involved could fail to make the same examination, to arrive at the s me result, and to publish that result in calm, judicious language, not as a matter of justice to Dr. Billings, or of criticism of Dr. Salmon, but as a positive duty to science. Regarded from this point of view, the report of that committee is the most disappointing document of the kind we have ever seen. This we may say without intimating, as Dr. Billings does, that its authors were under the control of the "Bureaucratic Whip."

The charges made by Dr. Billings, expressed in more conventional language than he employs himself, are that Dr. Salmon's bacteriological work is not expert, that he described a germ of swine plague when he had no such a germ in his possession, that he has printed assertions without scientific evidence to sustain them. He even states directly that one micro-organism as described by Dr. Salmon was evolved from the inner consciousness of that gentleman, and designates this procedure by a term to be found in the criminal code, and not usually employed in scientific discussion, however applicable it might be regarded in one sense to evidence, which Dr. Billings states to have been deliberately made up for the occasion.

Regarding the bacteriological question involved, we would be inclined to suspend judgment, were it not for two sets of facts First, the evasions of the "Board of Inquiry;" second, the gross logical errors and inherent contradictions of Salmon's writings. The latter are glaring. It would not have hurt Dr. Salmon's reputation had he, on being convinced of errors in his methods or conclusions, candidly acknowledged them, and accepted the corrections furnished by others and verified by himself. Men of a real rank in the scientific hierarchy like Virchow, Klein, Koch, and others,—and Dr. Billings here sets his opponent an example worth imitating—(No. III, pp. 50–53)—have not hesitated to acknowledge mistakes, such as the most honest and able investigators are liable to in experimental science, as well as in interpreting or quoting the results of others. But Dr. Salmon has preferred to gloss over the matter, and in order to sustain a claim to priority at the expense of another, and to conceal the fact that he adopted the corrections of others, has involved himself in a labyrinth of contradictions; the inevitable consequence of misrepresentations, be they intentional or unintentional. Which of the two classes Dr. Salmon's belongs to, no one who reads Dr. Billings' pamphlet and verifies its citations can for a moment doubt.—(*Journal Comp. Med.*, Vol. X, p. 399.)

Again the same reviewer says:

Regarding the bacteriological side of the question, the commission is equally unfortunate, and although the majority report claims that on the substantial point a dissenting minority report by Professor Bolton is in accord with their own, we do not think our readers will agree with them. Professor Bolton says:

"During my work as commissioner I have failed to meet with an epizootic which I am satisfied was what is termed "swine plague" in the bureau reports, though previous to my appointment on the board I studied one such outbreak. In this case, however, I directed my attention to the bacteriological questions exclusively, and I am therefore unable to pronounce on the difference in the pathological lesions in the two diseases. But I am not inclined to attach any great importance to these differences as set forth in the reports. The description otherwise I find correct and well stated. In my investigations as commissioner I have been able to find but one organism which in my opinion caused the outbreaks under examination, and that I regard as identical with the hog cholera germ described in the reports of the bureau, and I find the description therein given correct. As will be inferred from what has gone before, I feel sure that another organism, correctly described in the reports as the "swine plague germ," is found under circumstances which render it highly probable, if not certain, that it also causes disease. As to whether these two organisms are always present and operate together to cause disease, or whether the two are merely varieties of the same germ, must be decided by future investigation."

The main report insists that there are at least two widespread, distinct diseases which are caused by distinct micro-organisms. If this is what the committee calls agreeing, we would like to know what disagreement is. Who will be in doubt for a moment as to who, or what, induced Professor Bolton to write the last few lines of his report.

They are merely loop-holes for the escape of the man through whose hands the report went. Whether pity or courtesy were Professor Bolton's motives, he owed a higher duty to the country than to Dr. Salmon; we regret that he was not less ambiguous in expressing his convictions, although as contrasted with the majority report we have occasion to be grateful for even this much.

FOOLING THE SWINE BREEDERS.

There is a well-developed and continually growing belief among the great body of American swine breeders that they have been badly fooled recently—in short, they are commencing to look upon that meagre report of the so-called special commission on diseases of swine as a fiasco of the first water. 'Twas with not a little satisfaction that they regarded the appointment of the said commission. Were high-toned scientists to be relied upon the material of the board was surely satisfactory, and an appropriation of $30,000 for their expenses seemed ample to produce tangible results.

When the august body got into line of work December, 1888, swine breeders the country over were perfectly willing to wait until April 1 for the momentous findings expected. When April fool's day came they were, according to the custom of the day, fooled. The report came not. Premonitory rumblings of the approaching verdict were now and again heralded through the press. The mountain of scientific knowledge—the Parnassus of bacteriological learning—travailed in labor and brought forth—a mouse! The insignificance of the commission's production, so far as its intrinsic value to a long-suffering fraternity of breeders was concerned, was certainly in the proportion indicated, and any iota of value it may have contained was destroyed by the deceptive method of its publication.

A "boiling down" editorial bureau had been established at Washington and the commission's report was probably the initial work it edited. The press of the country looked for assistance from the said editorial bureau, and prepared itself implicitly to receive and publish as official and authentic the edited advance proofs the said bureau was paid to disseminate. Thus when about the second week of August a publication purporting to be the true and correct finding of the swine commission arrived from Washington, it was immediately published in full and commented on as the *in extenso* report. The *Farmer's Review* was not fooled, but in a short editorial outlined the weaknesses of the report, and since then has had no occasion to change its assertions. They will, however, bear considerable elaboration.

The Department of Agriculture editor, or mayhap some one more closely interested in the verdict of the commission, sent out the preliminary portion of the report with an explanatory introduction in which occcurred the following statement:

"We subjoin here the conclusions attached to the report of the commission, and signed, owing to the absence of Prof. Bolton, by Dr. Shakespeare and Prof. Burrill. Prof. Bolton, however, *furnishes a supplementary report practically confirming the report of his two colleagues.*"

The italics are our own, used to emphasize the assertions that Prof. Bolton's report confirmed that of the other commissioners. We shall see whether that was the case; but first let us ask why, believing this, Prof. Bolton's report was not sent out at the same time with the conclusions signed by Professors Shakespeare and Burrill? That it was not, leaves the swine breeder to accept the unpleasant inference that it was held back to give the conclusions the most favorable character as concerning Dr. Salmon and his Bureau of Animal Industry.

This subterfuge has, however, proved unavailing. It has been recoiled in hurtful strength upon those who employed it. The ire of the breeders and of the bulldozed press has been aroused since receiving the official report in full, of which Professor Bolton's contribution is perhaps the most important portion. The result must and certainly will be an unanimous demand for clean work in high places, for just value for public money spent, and for actual, honest, practical research in behalf of the inestimable important swine industry of this country.

As mentioned in a previous editorial in these columns, the commission's conclusions, as sent out by either the editorial bureau at Washington, Dr. Salmon, or the Bureau of Animal Industry, declare that there are two prevalent diseases of swine in this country, *i. e.*, "hog cholera" and Dr. Salmon's "swine plague." The former disease has in all conscience been bad enough. Our swine have had a sufficiently damaging character given them as regards disease. The industry has been hurt thereby. Germany, for instance, has barred out our pork, presumably on account of disease. The patriotic authorities at Washington surely forgot these things when they, with all their power of official advertising, solemnly declare that another serious disease is rife among our hogs. The breeders will thank them for this, doubtless. They will regard their $30,000 well spent in accomplishing such results. They will appreciate government aid in fostering their interests. They will admire the self-conceited confidence of a commission which, nothing daunted, proclaims to the wide world a new disease among our hogs without offering one solitary fact to substantiate the baneful assertion. Time will show what they really think.

And Prof. Bolton confirms this advertising of "swine plague," does he? Let us see. A comparison of the "conclusions" of Profs. Shakespeare and Burrill, with the independent minority report of

Prof. Bolton, *re* "swine plague," will enable the swine breeder to arrive at a correct conception of the truth of the matter.

CONCLUSIONS.

It is the opinion of the commission, based upon their own individual observations and examinations of the subject, that *there are at least two widespread epidemic diseases of hogs in this country* which are caused by *different micro-organisms*, but which have a clinical history and pathological lesions more or less similar and very difficult to distinguish without the aid of the microscope, and resort to bacteriological methods. * * So far as the knowledge and the observation of the commission go, one of the epidemic diseases, viz., that called by the authorities "swine plague," appears to be far less prevalent than the other, which has been named by them "hog cholera." The commission are further of the opinion that the disease called by the authorities at Washington "hog cholera" is caused by the specific action of a certain microbe named by them "the hog cholera germ," which has certain characteristics of form, size, movement, mode of growth in artificial cultures, and action upon certain lower animals, and taken together enable one to distinguish it from the other microbes which have been described from time to time by various authorities as present in swine disease; and that the descriptions of this microbe and its peculiarities, as set forth in recent annual reports of the Bureau of Animal Industry, are fairly accurate. The commission are of the opinion, *although to a less positive degree*, that the epidemic disease called by the bureau authorities "swine plague" has as its specific cause a certain microbe possessing characteristics which have been fairly well described in recent annual reports of the Bureau of Animal Industry, which distinguish it both biologically and pathologically from the first mentioned "germ of hog cholera."

PROF. BOLTON'S REPORT.

During my work as commissioner *I have failed to meet with an epizootic which I am satisfied was what is termed "swine plague"* in the bureau reports, though previous to my appointment on the board I studied one such outbreak. In this case, however, I directed my attention to the bacteriological questions exclusively, and I am therefore unable to pronounce on the difference in the pathological lesions in the two diseases. But *I am not inclined to attach any great importance to these differences* as set forth in the reports. The descriptions, otherwise, I find correct and well stated. In my investigations as commissioner *I have been able to find but one organism* which in my opinion caused the outbreaks under examination, and that I regard as identical with the hog cholera germ described in the reports of the bureau, but I find the descriptions therein given correct. As will be inferred from what has gone before, I feel sure that another organism, correctly described in the reports as the "swine plague germ," is found under circumstances which render it highly probable, if not certain, that it also causes disease. As to whether these two organisms are always present and operate together to cause disease, or whether the two are merely varieties of the same germ, must be decided by future investigation.

If these two reports corroborate each other we fail to see it, and we feel sure that breeders find themselves in the same position.

In addition to getting up this wholly unwarranted "swine plague" scare the commission seeks to throw cold water on legitimate and praiseworthy attempts to prevent "hog cholera" by inoculation with carefully prepared virus of the disease. Totally unable to destroy pigs thus inoculated, either by exposing them to an outbreak of cholera, or by feeding them with germ cultures, it warns breeders against inoculation as a preventive on the grounds that it may tend to spread the disease. The points made against inoculation are those long recognized and time and time again written about by Dr. Billings, who considers them of little moment if proper precautions are taken in carying out inoculation.

The fact that the inoculated pigs from Nebraska proved immune against hog cholera, as demonstrated by the commission, is practically the only valuable thing given to the public in return for the $30,000 expended.

The swine-breeding fraternity cares little or nothing about the *pros* and *cons* of learned squabbles about germs. It is interested in finding some method of preventing or alleviating the heavy annual loss from swine disease. Naturally enough it has looked to the Bureau of Animal Industry for help in this direction, but for the thousands upon thousands of dollars spent presumably in their interests they have reaped not a single *original* fact of real practical value in their everyday business. Time, indeed, that the work of such a bureau was thoroughly inquired into! Or, better still, that the government should recognize the fact that little but unintelligible bulletin publishing has hitherto been accomplished, and that the time has arrived when practical work is imperative.

Let the swine breeders of America unanimously demand such work. Let them hang no longer upon the words of false prophets or consent to be blindly led by the blind. Write off as lost the vast sums of money already expended in the research of swine diseases, and commence anew, and that now. The bureau has demonstrated its inability under its present head to serve the swine-breeding industry. Commissions are evidently a delusion and a snare. Other means and methods must be employed, and it is for the swine breeders of the country to demand the change in no uncertain voice.

Let the best men to be found the world over be employed forthwith, if possible, not as special commissioners, but each in his individual capacity, to investigate swine diseases in the different parts of this country. Let each man work for his own honor at the expense of the government, and offer them inducements for the discovery of disease cures or preventives. In this way good results will speedily be arrived at in place of voluminous and valueless bulletins offered to the long-suffering public in place of tangible duties performed at the public expense.

The *Farmers' Review* has come to the conclusion that swine breeders must put a stop to this fooling they have been subjected to, and demand and obtain the government aid, in regard to swine diseases, which is their right. This matter is one which demands united, energetic, enthusiastic agitation. Let the breeders strike while the iron is hot.—(*Farmers' Review, Chicago.*)

A QUESTION FOR THE COMMISSION.

"The Report of the United States Board of Inquiry Concerning Epizootic Deseases Among Swine"—the conclusions of which were submitted last week—has been received.

The letter of instruction to these gentlemen, signed by Commissioner Colman, was unmistakably drawn to direct the investigations of the scientists to the points at issue between the Department of Agriculture on the one hand, in the persons of Dr. Salmon and his assistants, and the Universities of Nebraska and Ohio on the other, as represented by Dr. Billings and Dr. Detmers. The commission was specifically charged to determine whether the work of the department was accurate and original, both of which points were in controversy raised by Drs. Billings and Detmers. That done, independent investigations were to be undertaken. The disputes between the doctors named who have been engaged in this line of investigation involved several points in bacteriological research—with which the *Gazette* and other laymen have nothing to do—and a vastly more important issue—whether there are one or two distinct swine plagues. The department affirmed there are two—the one "hog cholera," with its specific germ, and the seat of the disease in the intestines; the other, which is named "swine plague," with a distinct germ, the resultant disease of which finds lodgment in the lungs. The other investigators denied the existence of more than one disease and one germ, claiming that the work of this pest germ was at times manifest in the intestines, at others in the lungs, according to certain conditions.

Such was the chief subject of dispute. As stated in our last issue, on all points raised, the commission has determined in the department's favor. And yet there is an indefiniteness about its report that is very unsatisfactory. It declares that there are two widespread epidemic diseases of hogs, and that their description by the bureau are fairly accurate, "except it does not appear that 'hog cholera' of these reports can be said to have its special and exclusive seat in the digestive tract of the animal as distinct from the lungs." In view of all the circumstances this appears to be a very significant exception. "Swine plague"—the "new" disease—is pronounced far less prevalent than "hog cholera." After confirming the department's conclusions as to "hog cholera" and its germ, the commission turns its attention to the other diseases thus:

"The commission are also of the opinion, *although in a less positive degree*, that the epidemic disease called by the bureau authorities 'swine plague' has as its specific cause a certain microbe possessing characteristics which have been fairly well described in recent annual reports of the Bureau of Animal Industry, which distinguish it both biologically and pathologically from the first mentioned 'germ of hog cholera.'"

Is this the language in which science affirms its conclusions—"to a less positive degree"? Again, Dr. Bolton, the third member of the commission—who files a supplementary report—submits the following:

"During my work as commissioner I have failed to meet with an epizootic which I am satisfied was what is termed 'swine plague' in the bureau reports, though previous to my appointment on the board I studied one such outbreak. In this case, however, I directed my attention to the bacteriological questions exclusively, and I am therefore unable to pronounce on the difference in the pathological lesions in the two diseases. But I am not inclined to attach any great importance to these differences as set forth in the reports. The descriptions otherwise, I find correct and well stated. In my investigations as commissioner I have been able to find but one organism which, in my opinion, caused the outbreaks under examination, and that I regard as identical with the hog cholera germ described in the reports of the bureau, and I find the description therein given correct. As will be inferred from what has gone before, I feel sure that another organism, correctly described in the reports as the 'swine plague germ,' is found under circumstances which render it highly probable, if not certain, that it also causes disease. As to whether these two organisms are always present and operate together to cause disease, or whether the two are merely varieties of the same germ, must be decided by future investigation. The differences between them, as pointed out by the bureau, are sufficient to compel us to treat them as different germs, however perplexing it may seem that two micro-organisms are capable of producing such similar or, it may be, identical lesions."

Be it observed that although Dr. Bolton has not in all his investigation the country over as commissioner discovered one single case of this second disease—"swine plague," he yet "feels sure" that another germ is found under certain circumstances which render it "*highly probable*, if not certain," that it causes disease! On what evidence is he "sure" that it is "highly probable"? Is this again the language with which scientists seek to saddle a *second* swine plague on the country? The *Gazette* cannot be charged with hypercriticism of scientists, but it confesses it is unable to appreciate the force of Dr. Bolton's statement of the situation.

In fact the definiteness of the commission—"to a less positive degree"—as to the existence of a second swine plague, should have led it, the *Gazette* believes, to submit conclusive evidence to the country on this point. That a germ has been found which the Department believes is the cause of its "swine plague," does not admit of doubt; the question arises, however, is this the germ of merely a local, non-infectious, non-contagious disease, or is it a genuine *pest?* If the commission has answered this query conclusively, the *Gazette* has failed to observe it. "Hog cholera" is admittedly a plague; its germs, fresh from the blood or cultivated in artificial media, infallibly produce the disease when introduced into the systems of healthy swine. If "swine plague" is a pest its germ should likewise produce that disease with its specific lesions—but as we read the commission's conclusions, it can scarcely be said to have specific lesions. Now the commission had in its charge since February last, nearly twenty hogs, which in Nebraska had been inoculated against "hog cholera," and they withstood every attempt made by the commission to kill them by exposure to virulent natural outbreaks or by inoculation with excessive doses of the germ of "hog cholera." They were "hog cholera" proof, as the commission admits. But *these hogs were not tested with "swine plague" germs or exposure.* And the question which the *Gazette* respectfully asks the commission is, Why not? Here was the opportunity of all to demonstrate beyond all cavil the existence of a second *pest,* for if these hogs had succumbed to inoculations of "swine plague" germs, Dr. Billings would have been conclusively proved a charlatan, an ignoramus, and a fraud on his own chosen ground. He defied the commission to kill his inoculated hogs with any contagious or infectious swine disease; it could not kill them with "hog cholera"; if a second pest—"swine plague" exists, why was not its existence proved thus undeniably? If it kills swine as a *pest* these hogs should have succumbed to it as they had not been given immunity from *its* germs, and the commission was in possession of the material with which to infect and kill them if it could. Why was this not done?

The *Gazette* asks this question in all sincerity. It has the utmost respect for the scientific attainments of the members of this commission, and it cares not one picayune whether this or that investigator go down in the final determination of this much-discussed question; it is interested solely in learning the *truth,* but when it is asserted that two swine pests exist, the matter of the prevention of the disease by inoculation becomes necessarily so complicated as to be almost, if not quite impossible of attainment, and before that conclusion is definitely reached, swine breeders have a right to demand unconditioned proof. Why was it not offered, when the opportunity presented?—(*Breeders' Gazette, Aug. 21, '89.*)

HOG CHOLERA.

The following investigations were began in November of last year, and have been continued up to the present time. We were prompted to undertake them, not only on account of the scientific interest involved, but also on account of the news of many disastrous outbreaks of the disease in various parts of the state. In the course of our work we have visited many widely separated portions of the state, and encountered the disease in all its stages. In some herds the attack had just commenced; in others, we saw only cases of long standing, and, as will be seen elsewhere, the disease presents very different features in the two stages. We also found it in mild as well as in virulent form. In fact, in some cases very few animals died, whereas in others there were no recoveries. The majority of the outbreaks were of a very severe type. So we have been able to observe the disease in all its aspects.

We visited various portions of the state and took along all the best apparatus and appliances for our work. Material collected in this way was brought back to the laboratory and thoroughly examined. The germs which we found were cultivated and grown on suitable materials, and were tested upon mice, rabbits, and hogs. We obtained them from the spleen, liver, blood, and intestinal ulcers. In some cases we failed to get any germs at all, and although we obtained in other cases several different kinds, we have no reason to believe that more than one of them is concerned in the disease.

The only other organism deserving special attention is one which we regard as identical with the bacillus of hog cholera of Salmon in the report quoted above, which the reader is referred to for a complete description of the same.

We have not been able to produce the disease by subcutaneous inoculations of even quite large amounts (5 c. c.) of bouillon cultures of this bacillus, but we have succeeded by feeding 300–500 c. c. bouillon cultures in milk made alkaline with carbonate of soda.—(Bulletin No. 6, New Series II, July 1889, South Carolina Experiment Station.)

At a meeting of the United States Veterinary Medical Association, held at Washington, September, 1891, Dr. A. W. Clement made some remarks which bear on that report of the "Board of Inquiry," because "It has been my (C.'s) opportunity for the last three or four years of doing some work in that line myself, in association with Prof. Welch, of Johns Hopkins University. I might say in parenthesis, that my work in this line has nothing to do with my position as government inspector. As to what connection the organism (Salmon's swine plague germ) has with the lesions described in the reports of the bureau, is a question on which we all might not agree. Nevertheless the swine plague organism does cause trouble. *The trouble in hogs is, as a rule, in our experience, one of mixed infection. We have not had an opportunity of seeing an outbreak af swine plague (Salmon's) pure and simple.* We have found that it is very hard to say when swine plague (Salmon's) is present, that hog cholera (Salmon's) is absent, from the fact that swine plague kills [What, rabbits or hogs? B.] in a few hours, while hog cholera requires some days. If, then, an animal be killed and presents lesions in the intestines, as are generally supposed to be characteristic of hog cholera, the statement must be very carefully considered before it is made, that hog cholera is not present. We were thrown off our track during the earlier part of our investigations. *We found afterwards that hog cholera did exist in these animals that we thought had swine plague (Salmon's) pure and simple. I would simply say in a general way that from our investigations we have found Dr. Billings is right in certain other matters.*"—(*Journal of Comparative Medicine,* Vol. XII, p. 549.)

REVIEW OF THE SPECIAL REPORT ON THE CAUSES AND PREVENTION OF SWINE PLAGUE, PUBLISHED BY THE AUTHORITY OF THE SECRETARY OF AGRICULTURE, WASHINGTON, 1891.

For a number of years Dr. Salmon, chief of the Bureau of Animal Industry, and those associated with him in carrying out investigations regarding epizootic diseases of the pig, have maintained that in addition to hog cholera (swine fever) there is prevalent in the United States an infectious disease which they have been able to identify as the German Schweine-seuche.

This report gives a detailed account of the investigations and experiments that have led to the differentiation of the two diseases. It is a credit to the bureau as showing the energy and thoroughness with which its officers devote themselves to the elucidation of the obscure diseases of farm stock; *but we cannot say that a perusal of it has convinced us of the existence of Schweine-seuche on the American continent. Indeed, the impression left is rather that many, if not all, of the alleged outbreaks of that disease were instances of swine fever.* There has never been any doubt that hog cholera is identical with swine fever, *and we must emphatically refuse to accept as correct Dr. Smith's account of the morbid anatomy of the disease,* but to set forth the reasons for this dissent would occupy more space than can be spared here.

The report is commendable alike for its lucidity and the temperate language in which it is couched in dealing with controversive points, *and that is more than can be said of some of the literature on the same subject.*" [*Thanks, awfully. B.*]—(*Journal Comp. Path. and Therapeutics,* J. McFadyeau, Editor. Edinburgh and London. Vol. IV, p. 354.)

Mr. Rusk must know of all these verdicts against his publications and the conclusions of his investigators. He distinctly tells us that he has followed closely all the publications in Nebraska papers. These "reviews" have all been printed in the Lincoln, Neb., papers mentioned by Mr. Rusk. The old adage is again illustrated, "that there are none so blind as those who won't see." Mr. Rusk will not see the truth until the people of the country rise up and pull off the scales which the dextrous Salmon has drawn over his eyes.

How then could Mr. Rusk have written of me as he did in his letter to Mr. Paddock? What effect did he suppose that letter would have on the public? He knows now! It fell flat! No one believed a word of it as a statement of true facts. How can Mr. Rusk expect the people can respect a man who would be guilty of such a low device as to publish such a mass of falsehoods and misstatements as a public official? Surely our institutions are in danger if such men as Mr. Rusk has shown himself to be can only be had to fill our highest offices.

THE CHIEF OF THE BUREAU BEGS FOR SUPPORT.

Quite a number of such letters as the following have been sent me by friendly editors. They show a degradation in the public service, so far as the Agricultural Department is concerned, that is despisable beyond expression. Right never begs! Right demands until it gets recognition. Right stands upright. Right never bows the knee even to the mighty power of the press:

U. S. DEPARTMENT OF AGRICULTURE,
BUREAU OF ANIMAL INDUSTRY,
WASHINGTON, D. C., April 9, 1890.

Henry Wallace, Esq., Editor of the Homestead, Des Moines, Iowa.

DEAR SIR: For some weeks I have been receiving a copy of the *Homestead* marked "complimentary," for which I wish to return thanks, presuming that it comes from you. I have not been a reader of your paper before because it did not come to my desk, and, on account of the many duties devolving upon me, I seldom get time to consult the periodicals on file in our library. Mr. Hill, who has charge of the editorial division of the department, has frequently spoken to me about you, generally in connection with the attitude which your paper has assumed toward the Bureau of Animal Industry when discussing the subject of inoculation as a means of preventing hog cholera, and he has always referred to you in such complimentary terms that I venture to address you this personal letter in the hopes that any misunderstanding which you may have of our position may be explained, and that we may, if possible, obtain your good will even though our opinions may continue to differ.

The Bureau of Animal Industry is endeavoring to work for the best interest of the stock owners of the United States, and we are earnestly striving to furnish information which shall be as nearly as possible impartial and reliable. I take it for granted that the *Homestead* is trying to treat its readers in the same manner. We therefore meet on common ground; we are working in the same field, and if we differ in our opinions on certain questions, is that any reason why one should publicly refer to the other in such disrespectful terms as to tend to destroy his influence and usefulness in the work in which we are both interested? Is it possible that the farmers' cause can be promoted by those who are in the field wasting their energies and in destroying one another? I ask these questions frankly because I believe you are a reasonable man, broad enough and liberal enough not to be offended by them. Bear with me, if you please, while I make a few remarks inspired by the short article on page 3 of the *Homestead* of April 4. That I was not prejudiced against inoculation is shown by the fact that I was the first in the country to test it, and that before Dr. Billings began his investigations in Nebraska I had worked out the method of cultivating the virus, the proper dose to use for inoculation and the effect following this operation. And in the report of this bureau for 1886, pages 60 to 70, the details of the experiments are given so fully that any one who can make a culture of germs can repeat the experiment for himself. Against my own hopes I was compelled to decide that inoculation was not a satisfactory method of preventing hog cholera. No one could be more anxious than I was to offer some solution of the hog cholera problem to our farmers, and if I yielded my opinion to the inexorable logic of the facts, should I be censured for it? Admit, if you please, that I was wrong in my conclusion, would even that be sufficient to justify the language which some of the agricultural journals delight to hurl at me and at this bureau on account of the stand which I have taken?

The experiments which I have made have been planned to bring out the truth and the details have been published. That they show inoculation cannot be depended upon to protect from hog cholera under the conditions of these experiments, is plain. It is possible, however, as I have freely admitted, though hardly probable, that the conditions of exposure on farms are not so severe. In that case a degree of protection, which in our tests was insufficient, might ward off the disease in most cases on farms. But, surely, this ought to be demonstrated before farmers are advised to risk their animals and spend their money in having this operation performed, and this is what has not been done in any case by Dr. Billings. He has withheld the details of his tests, he has concealed his failures, and he has made claims which the facts do not justify. Take his experiment in Nebraska, where he had 1,000 hogs inoculated. Though he admitted at the time that about 400 of them afterwards died of hog cholera, he now says in his pamphlet that there was "a reported loss of only eleven out of the whole number." Can you recommend inoculation from that experiment in which 40 per cent of the hogs died of hog cholera? If not, where is the evidence which favors it?

Take the experiment in the article in the *Homestead*, to which I have already referred, where you say "This is the kind of proof required." Certainly you could not have scanned the details of that experiment very closely, because there was none of the conditions observed which would give reliable results. Consider simply the two inoculated hogs. I have it on good authority that they were the survivors of a lot of hogs which had been affected with cholera. [That is false! They were inoculated hogs and never had the natural disease. B.] Is it not presumable that in such a case, having resisted one outbreak, they might be expected to resist another without inoculation? In other words, when a herd of swine is affected the most susceptible animals die and those which live have more than an average power of resistance.

The proper way to make an experiment is to take a lot of twenty-five or fifty hogs which have not been exposed to the disease, inoculate half of them, then expose all in exactly the same manner to the contagion; keep all under the same conditions and note how much better the inoculated lot withstands the disease than the others. By making the kind of experiments which Dr. Billings reports one can prove anything, especially if he conceals his failures.

I have written much more than I intended to, but I trust you will not be bored by it. The subject is an important one and I feel sure that neither the *Homestead* nor any other paper can afford to either intentionally or unintentionally mislead its readers in regard to it. In my report on inoculation, of which a copy is mailed you, I have endeavored to place before the people the unvarnished facts according to the evidence now at hand, and before the conclusions therein expressed can be changed there must be additional experiments made which are properly conducted and which yield different results. I should be pleased to see some of the experiment stations take the matter up, and would assist them in so doing in any way in my power.

With the hope that this letter will not reach you on your "busy day," I am, very respectfully, D. E. SALMON.

THE EDITOR OF THE "HOMESTEAD" ANSWERS.

DES MOINES, IA., May 6, '90.

Prof. D. E. Salmon, Bureau of Animal Industry, Washington, D. C.

DEAR SIR: Your favor of April 9 came to the office while I was absent on a five weeks' business trip, and I take the first moment of leisure after my return to reply.

Allow me to thank you for the spirit and tone in which your letter is written, and to express the conviction that there is no need for men who differ honestly on certain measures of public policy to treat each other any other way than as gentlemen.

I may as well say to you frankly that the practical results of the investigations of the Bureau of Animal Industry with reference to hog cholera have not warranted any great hopes on the part of the swine growers of America. Possibly it is no fault of the bureau. Nevertheless, it accounts for the attitude which many of the farmers of the west sustain toward it and the favor with which they are inclined to receive the views and promises of Dr. Billings. It must be conceded that Dr. Billings has to a great extent the confidence of the leading agriculturists of the west, and what is more remarkable, the confidence of the leading stockmen of Nebraska. I attended their meeting in Lincoln in February and was more than surprised when they elected him, though not a resident of the state, president of the Nebraska Stock Breeders' Association for the coming year.

As I understand it, whether Billings' theory and practice of inoculation is correct or not, must be determined by the facts. When the Bureau of Animal Industry took the position that there was no remedy for hog cholera, and the government failed to take measures for stamping out the disease as they did the pleuro-pneumonia in cattle, it is not surprising that the farmers should turn for relief to Dr. Billings, who gave assurance that under certain conditions he could prevent the disease, and proposed to let the practical results of his works determine the correctness of his theory. The bitter warfare that has been made on Dr. Billings, and which is believed by his friends to come largely from the Bureau of Animal Industry, only intensifies the popular feeling on his behalf and it would seem to me that if the Bureau of Animal Industry, through its agents or employes, has been at the bottom of, or accessory to, this warfare, the best thing for all concerned is that it should cease. If Billings is either a fraud or an unbalanced enthusiast, time will very soon tell the story. If he is not, but has discovered a method by which, under certain circumstances and conditions, hog cholera can be prevented, then he is entitled to the credit.

Some of the statements you make differ from my understanding of the facts. It is conceded, I believe, by Dr. Billings' friends that 400 hogs inoculated were actually diseased, died of cholera, but it is not conceded that these 400 are part of the thousand to which you refer. I might mention other discrepancies, but have not leisure at this time.

I confess to you that the report of the special commission did more to prejudice me against the Bureau of Animal Industry than any other one thing. You will pardon me if I say that it seemed to be a whitewashing affair, sedulously, to all appearances as seen by an outsider, concealing facts and failing to make the investigations which it was appointed to make.

I have not, since my return, had time to go into the controversy between you and Dr. Billings fully. I stand ready to give the Bureau of Animal Industry credit for all the good work it actually performs and to insist, as I shall do in one of the coming issues of the *Homestead*, on the enlargement of its powers with reference to dealing with pluero-pneumonia in New York. I also stand ready to give Dr. Billings credit for any results which he may accomplish, the object in both cases being to secure as far as possible the financial well-being of the constituents of the *Homestead.*

Very truly, HENRY WALLACE,
Editor Homestead.

THE CHIEF OF THE BUREAU AGAIN BEGS FOR HELP.

U. S. DEPARTMENT OF AGRICULTURE,
BUREAU OF ANIMAL INDUSTRY,
WASHINGTON, D. C., January 28, 1890.

To the Editor of the Farmers' Review, Chicago, Ill.

SIR: In the October number of the *Journal of Comparative Medicine* was a review of recent swine disease literature, which in reality was an attack upon the U. S. Department of Agriculture, and upon those employed by it in the investigation of swine diseases. As I have seen references to this review in a number of agricultural journals I conclude that it was widely circulated by its authors, and for that reason I mail you to-day a copy of a statement which I have made in the January number of the same journal. This is sent for your information and you are invited to publish any extract from it that may seem to you of sufficient interest.

The Department of Agriculture has been striving for years to elucidate the question of swine diseases and has been, in the writer's opinion, eminently successful. The resources of the government have been brought to bear and have accomplished for the farmers what individual efforts never could have done. But this work has been criticised to an extent which is rarely experienced by those laboring for the public welfare in other fields of science. If these investigations have been intelligently and properly made, as we claim they have been, is it not a duty which the agricultural press owes to its constituents to sustain the department in its endeavors to have them continued rather than to discredit the work already done and thus give to congress the impression that the money has been wasted?

With accurate information I am confident the editors of the agricultural press will do justice to the careful and comprehenive researches of the Bureau of Animal Industry, and if the farmers of the country are given a plain statement of the facts, without distortion or misrepresentation, I am willing to rely on their unbiased judgment for the support which the department needs in its work. A portion of these facts will be found in the enclosed paper and others will be furnished from time to time as they are ready for publication.

Very respectfully, D. E. SALMON.

Does Secretary Rusk admit the necessity of such humility on the part of this department as this? Cannot it stand alone on its merits?

The next thing Mr. Rusk says immediately follows the words: "He published a pamphlet entitled 'Evidence Showing That the Report of the Board of Inquiry Was Fixed,'" etc.; then comes "*this disgraceful conduct, this use of the Experiment Station funds to carry on these bitter attacks against the bureau,*" etc.

Does Mr. Rusk mean to infer that that pamphlet on the Board of Inquiry was paid for out of the Experiment Station funds, or that I ever used such funds to publish a single attack on his department, except such as appear in the three bulletins published by the Station?

If he does, the strongest language a man can use is no more than his just due. Surely, the man has forgotten himself. When did he first send out the reports of that commission? By their date it must have been subsequent to August 1, 1889, and the first I knew of either of them was about August 12. I left Nebraska June 20, 1889, and was in Chicago in August, 1889. To be sure that pamphlet was printed by the State Journal Company, of Lincoln, but my money paid for it. Come, Mr. Rusk, come out here to Lincoln and go among the citizens and tell them that I have used the public money to publish attacks on you, and see how they will treat you. The following letter may be valuable to Mr. Rusk:

UNIVERSITY OF NEBRASKA,
OFFICE OF THE STEWARD AND SECRETARY,
LINCOLN, NEB., Feb. 15, 1892.

Dr. Frank S. Billings, City.

DEAR DOCTOR BILLINGS: Replying to your recent communication, I desire to say concerning the Agricultural Experiment Station funds that there has been no expenditure to my knowledge of the Agricultural Experiment Station funds since the organization of the said station, for printing, publishing, or distributing other than the regular bulletins of the station, which are required to be issued by law, and the necessary miscellaneous printing.

Yours truly, J. S. DALES, *Treasurer.*

Now, Mr. Rusk, how would you like to consume a little of your own medicine? You may find it bitter. We are considering a certain letter which you wrote Mr. Paddock in which it rather seems as if you were personally attacking me. Did congress ever authorize you to use the public funds to have that letter printed and scattered broadcast over the land. Answer that, please?

Did congress ever authorize you to send men over the country, and to publish pamphlets attacking the private business of a private citizen of the United States as you did me and my business while I was in Chicago from 1889 to 1891?

A congressional investigation might be in order, and would not be a bad thing either, as it would, if honest, expose the most rotten scandal, with Jere Rusk at the head of it, that has existed in the government since its inauguration.

Even a government secretary should carefully guard himself against that fell disease, *Macrocranialis giganticus.*

Next Mr. Rusk says:

Soon after his reappointment to Nebraska I received a letter from him asking that $5,000 of the funds of the bureau should be turned over to the use of the Experiment Station to be expended by him in his investigations, the condition being that the persons employed should be selected by him and that the Bureau of Animal Industry should have no connection with the work, and no control or supervision over the expenditure of the money.

That is not a true statement of the case by any means. What had the Bureau of Animal Industry to do with the whole matter? Is Mr. Rusk secretary of agriculture or is the chief of the Bureau of Animal Industry? Mr. Rusk should have a copy of that letter, I unfortunately have not, but it was about like this: First, I did not want and did not ask for any money whatever to aid me in my investigations. I made a kindly suggestion to Mr. Rusk which, had he been a true official, he would have endeavored to accept. I told him that

I was continually sent material from other states, or written to for advice regarding diseases in live stock, and offered him the free use of this laboratory for an investigator, who should not work for Nebraska, but for those other states, and stated that unitedly we thus could do much good. Of course I would select the man to come into my own laboratory, but with the approval of Mr. Rusk I supposed. My whole idea was to be of benefit to the farmers in other states. The man would have been under the control of the secretary of agriculture and not the chief of the Bureau of Animal Industry, which I innocently supposed the correct thing. The reports would have belonged to the department, and the man would have handled the money entirely subject to the secretary. I simply suggested $5,000 as enough to pay the salary and traveling expenses; all other expenses would have been provided for by this laboratory. That I would not recognize the bureau is true. Was the above either a criminal procedure or an insult to the secretary of agriculture? Some benefit to the farmers of the west might have resulted, but that would be contrary to the purposes of the bureau.

It is not known that during my stay in Chicago, and by every means in my power, both by letter and otherwise, I have asked Mr. Rusk to favor me with an interview, but in vain. Mr. Rusk would not be permitted to come into such dangerous society. The chief of the bureau dare not permit it.

But what is the National Agricultural Department for? Has the secretary of that department any right to discriminate between private citizens on personal grounds? During my time in Chicago I wrote Mr. Rusk the following letter, which any one can see was in the general interest of the swine breeders of the country. I wrote it so as to avoid making any mistake. I never received an answer to it, thoughI knew that it was received. I did receive the culture I wanted through the kind assistance of a gentleman who simply had to write and receive and forward it to me unopened:

CHICAGO, 6, 2, 90.

Hon. J. M. Rusk, Secretary, Washington, D. C.

MY DEAR SIR: After three years most exacting searching I have found what I take to be the germ of the so-called swine plague in connection with hog cholera and under just such circumstances as I should expect to find it, if any. It is very necessary that I should know without question that it is that organism, and after consultation with Mr. J. H. Saunders of the *Breeders' Gazette*, I take the only direct way open to me as a citizen of this country and a worker admittedly competent to judge. I ask you for a culture of the germ of disease called "swine plague," and no other, from the Bureau of Animal Industry, and that it be sent me carefully packed, by express, at my expense, and accompanied by a letter from yourself, by mail, asserting it to be the germ I ask for. As the germ has now been before the public three years and cultivations of the same have been sent to Boston, Philadelphia, and other places, there is no reason why it should not be placed in the hands of those desiring it to arrive at the truth regarding these matters. Trusting you will comply with my request, I am yours, FRANK S. BILLINGS.

Again, in order to do some very essential comparative work in connection with contagious pleuro-pneumonia and the corn-stalk disease in cattle, work that has never been done by the bureau, and of the utmost importance to the cattlemen of the west. I wrote the following letter to Mr. Rusk and again received no reply:

LINCOLN, NEB., 4, 30, 1892.

Hon. J. M. Rusk, Secretary Department of Agriculture, Washington, D. C.

DEAR SIR: My present work demands that I use some alcoholic specimens of genuine contagious pleuro-pneumonia material, and I know nowhere else to apply for it. Will you kindly instruct those having it to send me by express in small vials:

1. Freshest possible diseased lung.
2. Tissue adjacent to moderately diseased lung.
3. Moderately diseased tissue.
4. Chronic diseased tissue.
5. Tissue in process of destruction.

Respectfully yours, FRANK S. BILLINGS.

These requests on the secretary of agriculture were not made for my own benefit or for my own personal use in any sense of the word. Both letters were for material to aid me in investigations of the most intimate kind to the welfare of the stock breeders of the west, and, as I understand it, that is one of the purposes for which the Agricultural Department at Washington was created. Suppose Mr. Rusk, or even the chief of the Bureau of Animal Industry, should write me for cultures of any germs that we might have in this laboratory, does any one suppose for a moment, that such a request would be refused merely because my relations with either were not as cordial as they should be? Does any one suppose that did I treat them as they have me, that a letter of remonstrance would not be soon sent to the chancellor of the University, and in all probability to the public press? I should not think of refusing such a request, because, this, like that of the Bureau of Animal Industry, is a public laboratory, and all in it, as well as what goes on in it, is the property of the people. They are carrying things in a far too "high handed" manner in the Department of Agriculture for officers who are mere delegates of the people under a republican form of government. Is it not time that the farmers of the west make themselves heard and teach these people that their position is to serve the people and not to follow their own selfish impulses? *Is it not about time that the people once again declare the "declaration of independence" over again, but this time against the political kings in their own couutry who have robbed them of the rights their fathers won for them on many a bloody field? Is it not time that the "constitution" be once more ratified and that the people again declare that this is a government "of the people, by the people, for the people," and not one of a nation of disfranchised slaves by a band of political tyrants and traitors for their own benefit?*

TWO OPEN LETTERS

PUBLISHED IN

WESTERN RESOURCES

WHICH FAILED TO PLEASE THE

SECRETARY OF AGRICULTURE.

THE HON. MR. RUSK DOES NOT LIKE OPEN LETTERS.

Having utterly failed in obtaining any consideration of polite letters written direct to the secretary of agriculture, and being desirous to force the issue, if possible, so as to end this interminable controversy, which I have only entered on and continued as a public duty, I next had recourse to several open letters published in the columns of *Western Resources* of August 20 and October 10, 1891, which were as follows. As, with this publication, it is my intention to permanently close the discussion with the Agricultural Department, the public must pardon me if I sum up all the evidence at my command in full:

PATHO-BIOLOGICAL LABORATORY, STATE UNIVERSITY,
LINCOLN, NEBRASKA, August 10, 1891.

Hon. J. M. Rusk, Secretary Department of Agriculture, Washington, D. C.

DEAR SIR: On June 26 last I mailed you a registered letter, and on July 4 received a card signed for you by "Thos. J. Ray," thus indicating that you had received said letter. In that letter, among other matters, I requested you to send Dr. F. E. Parsons here to personally investigate the question of inoculation against swine plague in connection with my assistant, as two unitiated persons, and offered you the free use of everything we had. I stated to you that Dr. Parsons had made a very favorable impression among the stock breeders of the state, and that we considered him an honest man. That letter has not been answered, and occupying the position I do—possessing the trust of the people of the state, as probably no other man in like position does in the world wherever he may be—I beg to say that you may not realize that your conduct in this matter is not in accord with the spirit of the promises made in your name by Dr. Parsons, at Beatrice, February last, who said: "You wanted to be in touch with all breeders and agriculturists," or words to that effect. Now, again, in the name of the breeders and farmers of Nebraska, and as their selected and elected representative, I ask you to send Dr. Parsons here on the same terms as before. If you want to know how long it will take I will tell you that in three months he shall himself have inoculated hogs successfully, and I hope have seen many hundred done, and that we will all do our utmost to give him every opportunity to test the matter. Our Mr. Paddock and other very influential men (who are as much friends of mine as friends of your department) are all anxious for me to use every inducement to bring this about; and I will respect their desire quietly and gentlemanly if I can, bitterly and determinedly if I must, for, I tell you, that not even in yourself, not even in themselves individually and collectively, have the breeders and farmers so true and determined a friend as I have been, and am, and mean to be. I told you long ago you were being misled, and I now tell you you are dangerously so. They tell me that you are an honest and clear-headed man; if you are, then as Mr. Rusk, farmer, and not Hon. J. M. Rusk, I ask you to turn to the last report of the Department of Agriculture, and read what the Hon. Mr. Rusk says to Mr. Rusk, farmer, and then see what the latter thinks of it:

SWINE DISEASES.

"An experiment to test the value of subcutaneous injections of hog cholora bacilli as a means of preventing hog cholera.

"In the report of 1889, page 87, it was stated that an experiment was in progress which we hoped would be a final test as to the practical value of subcutaneous injections of cultures of hog cholera bacilli, in making swine insusceptible to the virus of hog cholera. The first tests in this direction were made at the Experiment Station early in 1886, soon after the hog cholera bacillus had been discovered."

How, sir, you as secretary could tell yourself as farmer that "the first test in this direction (of inoculation) were made early in 1886, soon after the hog cholera bacillus had been discovered," beats me? Of course neither as secretary or farmer do you know that statement to be strictly and reliably correct. You dare not and cannot take oath, as secretary, to Mr. Rusk, farmer, that the evidence furnished is all that can reasonably be required to decide a scientific question of this kind (although you did take oath as secretary, to be true to Jere Rusk, farmer, and all his co-laborers), because the same persons now in your employ once told you, as farmer, something very different; and, if in your regular capacity as farmer, you will turn to a report of your department for 1883, p. 57, you will see that, as secretary, you are on the point of both horns of that dung-hill steer, "dilemma," and if you try to get off one horn you will fall into the ditch, and if off the other you put one of your predecessors into the ditch. But you do not see that the only way out is to get off both horns, and pitch that irresponsible crowd who are deceiving you into the ditch of oblivion. Now, let us turn to the report of 1883, and read:

"Our investigations show that swine plague is a non-recurrent fever, and that the germs might be cultivated; they have even proved that these germs may be made to lose their virulent qualities and produce a mild affection. Surely we have here sufficient evidence to show that a reliable vaccine might easily be prepared if we carried our investigations but a little way further." "M. Pasteur has confirmed our investigations and shown that the disease is produced by a micrococcus, and promises a vaccine by spring."

Now, sir, as you, Secretary Rusk, tell Farmer Rusk that the hog cholera bacillus was not discovered until 1886, "what must Farmer Rusk think of those other investigations" which promised him so much in 1883, and of all those costly reports, 1880 to 1886, which you now tell him to be unquestionably false, seeing, as you say, "that the first tests in this direction," of preventive inoculation, were not made "until early in 1886"? We want these questions answered.

Again, sir: The following interesting letter recently emanated from your department, and is in fact, as well as in spirit, most emphatically contradicted by other emanations from the same source:

"U. S. Department of Agriculture,
"Bureau of Animal Industry,
"Washington, D. C., June 11, 1891.

"*Mr. James Baynes, Editor and Publisher American Swineherd, 113 Adams Street, Chicago, Ill.*

"Dear Sir: I am in receipt of your favor of the 29th ult., which has been held a few days for consideration. In it you ask a number of questions in regard to inoculation as a preventive of swine plague and the probability of disseminating that disease by the practicing of inoculation. The subject is a complicated one and I could not give you my views without writing you an extended letter. I mail you, therefore, a copy of a bulletin published by this bureau about a year ago on this subject:

"In a general way I would say in this connection that much of the difference of opinion arises from the fact that there are two diseases, one of which we have called swine plague and the other hog cholera, but which are not discriminated between by the public at large, and which are consequently referred to as hog cholera or swine plague without any idea as to which disease is under consideration. A hog that is inoculated for swine plague is not in the least protected from hog cholera, or *vice versa*. The inoculation that has been practiced in the west has been with the virus of hog cholera, and, as used by the parties who have introduced this practice, it produces little if any immunity from the disease.

"As to a second question, I would say that I don't believe that inoculation could be safely trusted to the use of the average farmer, or for that matter the average veterinarian, as a practical preventive of hog cholera; in fact, I do not consider it a practical preventive under any circumstances.

"Replying to your third question, I would say that I have no means of knowing whether the wide prevalence of swine disease was or was not due to the dissemination of the disease by the many attempts at prevention by inoculation. I have no doubt but that the disease may be spread in this way, and if, as is intimated in your question, there were many attempts made at prevention by inoculation, this may account for the unusual dissemination of the disease in the sections where inoculation was resorted to.

"Referring you for the reasons of my opinion to the bulletin above mentioned. Very respectfully,

"D. E. SALMON, *Chief of Bureau.*"

Take the last paragraph first. It is here inferred that I spread cholera over the country because of my method, which is condemned as so terribly dangerous. My dear sir, do you know that you, Hon. Mr. Rusk, secretary, recommend that method to Mr. Rusk, farmer, and every other farmer in the United States in your late report (which only antedated that letter by a short time) as the cheapest and simplest method that can be devised? Perhaps you do not believe you did? Then read, please, pp. 110, 111:

"The method of subcutaneous injections of culture liquids containing hog cholera bacilli, while on the one hand fraught with the possible danger of scattering disease germs where they do not originally exist, is nevertheless the simplest and cheapest method that can be devised for the vaccination of animals; these qualities of simplicity and cheapness are of vital importance in a question which has only a commercial aspect." That is straight, is it not?

As to that two-plague question, it is equal to the others, and all false. You should know that no impression has been made in Europe with that stuff, and the universal verdict is against you. I now know more than I did on the subject, and will soon be able to knock all the plague there is in it to pieces. To-day there is no conclusive proof of even two diseases, as no person on earth conversant with the entire history of the investigations done in your department will believe a statement of facts emanating from there unless known to be such by his own experiences. It is told in this letter to *The American Swineherd* that farmers dare not inoculate because of their inability to tell one disease from the other, but you forget that when you went, by proxy, before the National Swine Breeders' Association you said: "There is the genuine hog cholera, a disease widely distributed, and probably productive of the greater part of the losses which fall upon the hog raiser." (See the report of the association, p. 13, 1889.) Now, if the method employed by me is the cheapest and simplest, notwithstanding the remarkable financiering of your department, and capable of preventing "the greater part of the loss which falls on the hog raisers," will you kindly tell us why it is you are so loth to send any one here to see how it is done, and work it out for himself? It is no argument to assert that "we cannot do it, and do not believe any one can," or that "because you cannot, any one who succeeds is a fraud," which is about what you have said of me to the entire world, but fortunately it has been a boomerang that it is still reacting. Because I am not equal to Edison's work, or that of many other men, is that a reason why I should brand them as failures or frauds? Let me tell you something that you do not seem to be able to see. Your employes dare not succeed, for, as admitted in this very report, 1890, to succeed they must use those terrible "fluid cultures" which, though "fraught with possible danger" (only possible you see, not probable) "of spreading the disease, still is the simplest and cheapest method." I tell you these men cannot and dare not succeed, for that would brand them as thoroughly incompetent and irresponsible investigators before the whole country, and bury them so deep that even your stentorian voice could not raise them from the grave of oblivion. It would confirm all my statements. One word more. As said before, I mean business. Hence, in the name of the breeders and farmers of the state of Nebraska and of every stock raising state, I again ask you will you or not send Dr. Parsons out here to work as suggested, independent of me, for a period of three months? The farmers desire the truth, whether others do or not. Are you also afraid of the truth, Mr. Rusk? You may not like the tone of this letter, but I assure you it is written in the kindliest spirit (and in your own interest), but more particularly in the interest of the farmers and breeders of the country.) Yours most truly,

FRANK S. BILLINGS, *Director*.

LINCOLN, NEB., October 4, '91.

MY DEAR SIR: Some six weeks since I addressed you a private letter, and later, in the columns of *Western Resources*, an open one, asking you to send Dr. F. E. Parsons here, not to investigate our work, but to observe everything we did in the process of inoculating against swine plague. Still later I urged you again to send Dr. Parsons, hoping he might come in time and learn enough to be able to take charge of the inoculation during my absence, as I take my vacation in October and November. The only reply I received was that you would consider the matter on the return of Dr. Salmon from Europe. This is a question between you and the farmers with which Salmon has nothing to do. Yesterday a dispatch appeared in the Lincoln *State Journal*, coming from the acting chief of the Bureau of Animal Industry—a good man, a competent one in that position, and an honest one also—Dr. C. B. Michenor, whom I suppose spoke authoritatively. He says you are looking about for a scientist that would be impartial both ways, and cannot find one, but thought one might be found by next summer. Here is what he says as reported:

"Dr. Michenor, who was in charge of the bureau to-day, said that the department had not had any success in its experiments in inoculating swine, but he said: 'It is possible that Dr. Billings has discovered a method which will prove effective. It is not true,' he said, 'that the department takes no interest in this matter. On the contrary, we have been trying to find the right man to send out to Lincoln for the purpose of inquiring into the matter. The difficulty is in finding the proper man for the work. It would scarcely do to send a man who might be prejudiced in advance for or against Dr. Billings. We have no inspector here who combines the necessary knowledge of swine diseases with the requisite familiarity without use of the microscope. Consequently it is necessary to get some one from outside the department. Thus far we have not been able to secure the services of the right man. It is not likely that we shall be able to do so this year, but next summer one will undoubtedly be sent to Lincoln to represent the department, and to make a thorough investigation into the process by which Dr. Billings is said to prevent the diseases."

Now, though I fear nothing nor anybody, a scientist is just what we do not want and will not accept. I would not let a single one now known to me in this country into my laboratory and observe my methods, for I do not personally know one engaged (or dabbling—the latter mostly) in this kind of work whom I look upon as an honest and unprejudiced man. That there are such I do not doubt, but I fail to have the honor of their acquaintance. You may not know it, but this state and the swine breeders of the west had all they wanted of that kind in the "Swine Plague Commission" that visited here in the winter of 1888–9. I have written it elsewhere, but you may not know it, that these men could not be induced to make an investigation while here, and only when I sent my own servant to drive one of them into the country could they be made to examine one sick hog on their own account. All they seemed to want to do was to have me work and see how they liked it, in which you may be sure they were not fully gratified. They went to Columbus, Ohio, and did not examine even one hog there. Then they went home, and two of them, at least, wrote what they were told to, for they did not even report on the examination of the one hog they did make an autopsy on here and take cultures from. This was probably because they failed to find the hypothetical "swine plague" germ of your bureau in it. Had they found it the world would have been told of it in no very uncertain language. They did happen to see me make some half-dozen autopsies on hogs which I had purposely killed by inoculation as a control against a bunch of inoculated ones, which they knew all about. I killed fifteen of the healthy test hogs, but the same dose of the same virus failed to make even one of the preventive inoculated ones sick an iota. This is a matter of history out here to which men of undoubted honesty can and do testify. But more, these tests were going on as that commission came out here, and I insisted that they personally take cultures from the hogs, which they did, and I made and dictated the autopsies which one of them wrote out, the others watching me. I insisted only that they should give me one duplicate culture from each organ of each hog, which they made, to send to Prof. Welch, of Johns Hopkins, who wrote me that they were nearly every one strictly pure cultures of real swine plague (hog cholera) germ. You can depend that he sought for the hypothetical swine plague germ, as did your commission. Why then did not they report these autopsies and the results of their examinations of their own cultures? Because they did not find that second germ! That is all! They were sent out here to investigate and report on what they saw, and were called, and were really supposed to be, scientists, and my experience with German investigators had taught me to look upon all scientists as honest men. Were these men honest when they did not report on what they did find, and did report on what they did not find, viz., a lack of exact technique in my methods? It is because these men are known to be dishonest and not to have reported on what they did find, and because it is known here that they would not go out and investigate as honest men should, that the stock breeders of this state want no more of that kind sent here. I speak with the authority of the best breeders of the state, of the State Board of Agriculture, and the Regents of the University when I make that assertion.

When I asked you to send Dr. Parsons I was supported by the same authorities who unitedly have control of me and my work here. But more, I never do a thing without a purpose. Personally I like Dr. Parsons, and that is probably one reason that he has not been sent. Again, our breeders who met him all liked him. You must both like him and believe him an honest man or you would not trust him. When I asked you to send Parsons the people here said: "Rusk won't do it. Salmon won't let him." It certainly looks that way to us now. Cannot you trust a man known to you for so many years and selected by you for your especial work? I want to see more workers in the country, and, liking the doctor, I expected that in the time he would be obliged to stay here he would learn enough to be able to go on and become an investigator, for when I like a man I am willing to teach him all I know and to work unceasingly to aid him. That was my chief reason for asking for him. We do not care an iota about an investigation. We can paddle our own canoe and hoe our own corn. In the interests of our brother farmers in other states we simply offered you the opportunity to see our work, which offer you seem loth to accept. You may be very sure that a person whom I do not want cannot come here so long as I have charge of this laboratory, and least of all a "scientist," for the reasons above given.

I am trying to write with studied respect, so do not be offended at my plain speaking. It is not necessary that a "scientist" should study this question. You have some of that species in your employ who claim to have done so for ten years and, notwithstanding certain unfounded promises in 1883, have still failed to do one single thing to benefit the swine growers of the west, unless "benefit" can be seen in the utterly unfounded announcement of a second "widespread epidemic disease among the hogs of this country." One word on that. It is claimed that my methods are unscientific, or else I should have found that germ, and hence my inoculation is dangerous. My dear sir, I have furnished the virus to inoculate many thousand hogs—the farmers having done the work—and I challenge you to find one man who will assert to-day that inoculation ever made a hog or pig seriously ill? I even dare refer you to the notorious case at Surprise, Butler county, Nebraska. Try those same men to-day, and see if they will say what they erroneously said in 1889? I have never seen one of them in person, nor written them. Now, if my methods are so bad, and that second germ so dangerous, why has not a farmer who has used it killed one pig by inoculation? I have found your germ. I found it in three outbreaks out of 150 or so in Illinois during my absence in Chicago, and have since found it here—always mixed with hog cholera—and also where no disease was present and in hogs having another disease which I am now studying. I have found it in the blood of cholera sick hogs—but in far less numbers than the real germ where were neither intestinal nor lung lesions of any moment whatever. I have found it when it would kill rabbits in less than twenty hours and from pure cultures of that same germ have inoculated virus with it, and the real germ and both have grown together and inoculated nearly 2,000 hogs and little pigs from two weeks old upwards, the farmers doing the work, and not one single pig has been made seriously sick. The thing may be terribly dangerous in Washington, but if so it is not the same thing I have found in Illinois and Nebraska. But this coincides, unfortunately, with the general results reported by your own people. These statements are all facts of exact record, 2,700 pigs and hogs have been inoculated in Nebraska from August 14th to date, with no injury and no loss thus far. I expect some, for the greater the number done the greater the chances for error, and the farmers sometimes neglect to use the virus as directed.

Now to the scientist once more. You are simply a practical every day man, and those who know me well, know that I am built about that way myself. As such, we both know, and all the world well knows, that a man, especially a scientist, can be too much of a stickler for a theoretic method of work, which for certain purposes is all right, but for others may entirely destroy or nullify the very practical purpose which should result, or better, is wished for, from a certain series of investigations. Now I am not such a hard and fast scientific investigator. I go for the practical point, and that is all I am engaged for, though I can do exact work when necessary. But here the exact scientific method has necessarily been made supplementary, though co-equal with the practical. You may be very sure it has not been neglected. Now one reason why your men have failed in making any advance in preventive inoculation is self-evident from their reports. They are too punctillious as to the theoretic exactness of their methods a la Robert Koch. No disrespect is meant to Koch. To demonstrate scientifically that a given germ is the actual cause of a given disease all the finical exactness possible is necessary, but where practical results only are desired, as in preventive inoculation, that point may be rendered impossible through the prolonged unnatural and artificial treatment of the germ necessary to obtain pure cultures. In spore-bearing germs this is not so, but in swine plague it is the case. In my investigations here I could not afford to waste time seeking for all the germs mixed up with the swine plague germ in certain kind of outbreaks. The people would not have the patience to wait and the work would have been stopped in disgust. Look at the way the farmers of the west view your bureau. The thing to do was to find out not only the kind of an outbreak (the point is, the nature and lay of the land and how the hogs are kept), and then the organ from which one could with the greatest certainty obtain the germ as uncontaminated with other germs as possible. These points were soon established. Having done this, the next thing to do was to discover the kind of an outbreak from which the germ would give the most reliable degree of protection, and a third was, the time in the outbreak at which to derive the germ. This will explain a mystery which seems to trouble your people very much, and that is, why I do not mention rabbits in my experiments. My dear sir, I am studying swine plague in swine and not in rabbits, and swine or pigs are much cheaper than rabbits in this "section of God's country." Again, I found, to my cost though, that results in rabbits are not only valueless but apt to be absolutely misleading so far as they have any practical value in the study of the swine plague. In fact I make it a rule to use the same species of animals in which a disease naturally occurs, for my experiments, as far as I possibly can. This may not be scientific, but it is decidedly practical. It has won so far at any rate.

The point we are aiming at, and have most decidedly obtained, is a method of obtaining and preparing the virus so simple, easy, and reliable that any of our farmers can do the whole thing themselves after a few hours' instruction. This I say can be done and with a reliability so certain that failure is almost entirely ruled out.

Does it require a scientist to investigate what a plain farmer can do?

Now suppose you send your scientist, one of the stickler, punctillious kind, what on earth could we do with him?

First, he would say, "Why, how do you know your virus is surely composed of the swine plague germs and no others, if you do not make plates, cultivate on potatoes, in milk, and half a dozen other tests, no one of which is absolutely so practically reliable as the plain, every day fact that the culture came from a cholera sick hog?

Again, while your scientist is carrying his cultures through all these methods and going through all this exact scientific detail, he is removing his germ farther and farther from its natural conditions, and you may be sure he is losing something that he will never find again in that culture.

What is that?

The preventive qualities of the germ.

As sure as you occupy the position you do that power will be lost, and at present can never be regained. We can retain the virulence in cultures indefinitely, but cannot preserve the prevention. The rule is, that in order to obtain reliable prevention the first culture from the sick hog can be absolutely depended on, and that every succeeding generation is less and less reliable, and that even the third and fourth should not be used if much of anything depends on it. That is, transfers being made once a week. In four weeks the prevention is practically lost. Naturally the right kind of an outbreak must be selected and the hog condemned at the right period of the disease.

These points have all been settled here, and this is what we have been at work on, instead of wasting our time seeing how many tails a germ had or how many "stripes it had on its belly," or how it would grow on green cheese direct from the moon, or some other bacteriological demi-nonsense. Pretty work, it's true, scientific it even may be, but its practicability is very questionable.

It may be said "that the farmers, not being scientists, will not always get pure cultures in virus prepared by themselves."

That is true scientifically, but absolutely absurd practically.

Generally, an experience of five years in the west tells me that they will get pure cultures. For instance, I can honestly say that out of about 150 hogs, each one from an independent outbreak, I obtained absolutely pure cultures in all but three, during the time I was in Chicago. What is more to the purpose, rabbit controls were used in every case. The heart's blood of the hog was only used to obtain the virus from. But suppose there is an occasional extra germ in the blood, is there any harm or danger in that fact if the swine plague germ predominates, as it surely will?

We all know that the danger of the other germs getting into the blood of the hog is very small indeed, by inoculation, and even if a stray bureau germ gets in I, for one, do not fear the result. What of that?

Let us stop all nonsense and get down to actual facts. Do you know of a single successful virus that is being used to-day that can be said to be pure culture of the germs of the disease which it prevents?

Take vaccination, for instance. I believe the real germ has been discovered though the facts have not been published in detail. For a century vaccination has been practiced, and with no precautions as to germs; for no mortal knew, and but a select few to-day know, which or what is the germ of vaccina, yet on every point used are many varieties of germs which do no harm in general. I have seen cultures from points of a great variety of germs, not one of which when isolated proved to be the germ of vaccina, produce the most characteristic vesicles, as is the case in vaccination as used everywhere. Why? Because the real germ was among them. The same results, but no better, have been produced with the real germ when isolated and used by itself. The vaccine vesicles of the calf are exposed to all manner of pollutions, and every point used contains at least ten adventitious germs to one that will be found in a culture of swine plague virus properly selected and taken direct from the heart's blood.

No one doubts that Pasteur's inoculation against rouget worked and will work, and those who are acquainted with the facts know that Pasteur used it several years without ever knowing the actual germ, and probably without ever having seen it. Still, it worked! The real germ was first discovered in Pasteur's virus, and found to be a delicate rod by Schütz in Germany. Pasteur had said it was a dipplo-coccus (a figure 8), but was mistaken. Schütz found six or seven other germs in Pasteur's virus against rouget.

And still it worked!

None of these preventive inoculation viruses have worked any better, and few as uniformly well as that against swine plague in my hands and in the hands of our western farmers.

That is all you want. We want no technical scientific nonsense, for such unfortunately exists, if we have a plain, clear, square method which every farmer can apply himself.

That we claim to have, and that we are demonstrating as fast as possible. Scientific painfulness is very liable to destroy its value. The scientific exactness lies in the selection of the right hog at the right time, which any farmer can do, and not in the technical details of laboratory exactness for exact demonstrative purposes. The time occupied in carrying such out will certainly nullify every preventive result. We do not want your scientist and will not have him, as we have no time to bother with such, but if Dr. Parsons cannot be sent, should some every day, common, ordinary, practical swine breeder visit us, we shall give him a royal welcome, though we should not bother him much with a microscope, but rather send him down to our friend Walker's, and a number of other men of the same stripe, and let him see how a man of his own calling inoculates his own hogs.

Why not visit Nebraska yourself?

"Next summer!" Why delay that long? The farmers of the west will certainly lose millions of dollars from swine plague ere that time arrives. Procrastination is not thief of time alone, it's killing our farmers. Then we have very little disease in the summer months, and it is "between hay and grass," the spring pigs will be grown and the fall ones not born, and the farmers so busy in harvest that little or no inoculation will be done. You have been a farmer and should "know how it is yourself" at that season of the year. Again, by that time inoculation will be entirely out of my hands and in those of the farmer, supported, perhaps, if necessary, by some competent young woman, who will have entire charge of it in the laboratory. There are no mysterious methods to be studied. It's all plain sailing on an open sea. Do not you think it would be a good idea to order your workers to find out inoculation before spring, or procure others who can? It took me just four months to demonstrate its practicability from the time I began to work here. It can be done. Why not put the testing on where it belongs and not where it is about over? That is what the farmers are looking upon you to do. Take a glance into Ohio. I see they are succeeding there also. In fact, Mr. Rusk, remove the mote from your Washington eye. You can't find any motes in the western one. In sporting vernacular, the people here think it is about time for the Bureau of Animal Industry to either "put up something of value as a testimonial of so many years' existence, or shut up" as to what others are really doing.

Yours most truly, FRANK S. BILLINGS,
Director Patho-Biological Laboratory, State University of Nebraska.

Reader, is there anything so terribly impolite or wrong in those two letters to Mr. Rusk, even though he be the "Honorable Secretary of Agriculture of the United States of America"? Of course they might be considered audacious impudence had they been written to King William or Prince Bismarck, of Germany, but one thing is sure then, there would have been an honest investigation and the sinner, who ever he is, would have been *in quod* by this time. The dishonest work which has been going on in the Agricultural Department at Washington ever since 1880, by and through the influence of the chief of the Bureau of Animal Industry, could not continue for six months in Germany. In fact, it would never have been begun, for German investigators are not built that way. They are men of honor and scientific reliability. They are not tricky and irresponsible politicians. Why should I not have written Mr. Rusk as I did? Am I not a citizen of the United States? Am I not admitted by the entire scientific world to be competent to express an opinion on the questions in dispute? Have I not then even more right to express my opinion, in public if I choose, than the editors of daily papers on questions like this? This government is not, theoretically, an imperial autocracy, though it begins to look as if the secretary of agriculture assumed that it is so, so far as he and his department is concerned. I did not know then, and do not now, that Mr. Rusk is secretary of the Department of Agriculture "by the Grace of God" as William III audaciously asserts of himself that he is King of Prusia and Emperor of Germany?

It is but natural that Mr. Rusk should not have been pleased with such plain spoken letters. They demonstrate the perfidy of a department under his charge altogether too acutely to be agreeable. Of these two letters he says in his letter to Senator Paddock:

Under date of October 9, '91, he published an open letter in which he quoted from department reports, garbling his extracts, and misrepresenting the position of the writer in order to deceive his readers.

Every word of that accusation is emphatically false. Why not show up the "garbled extracts." "As to a desire to mislead my readers" such a thought never entered my head. I am not a scoundrel, Mr. Rusk, nor is my head so wanting in the level qualities that I am not very cautious in all my quotations. I have never intentionally quoted for mere effect in my life, but to place the square truth before the world. When I published my Swine Plague Report, 1888, I took pains to read the entire manuscript over with a disinterested person, who had all the department reports and other documents quoted from at hand, and not one was objected to as incomplete or for mere effect. Another and entirely disinterested person reviewed both the work of the department and myself most critically and expressed the following on this point:

We have compared the extracts cited by Dr. Billings from the reports of the Department of Agriculture, line by line and word for word; we have carefully examined the plates accompanying these reports to which he refers, and we have also read the entire context from which he quotes, in order to avoid any possible bias which may follow from reading disconnected sentences. As a result we are enabled to say

7

that all of Dr. Billings' charges, sweeping and severe as they are, are true and just, and we cannot understand how the committee appointed to investigate the special question involved could fail to make the same examination, to arrive at the same result, and to publish that result in calm, judicious language, not as a matter of justice to Dr. Billings, or of criticism of Dr. Salmon, but as a positive duty to science. Regarded from this point of view, the report of that committee is the most disappointing document of the kind we have ever seen. This we may say without intimating, as Dr. Billings does, that its authors were under the control of the 'Bureaucratic Whip.'"

Still another independent and personally known to me as a very cautious critique has expressed himself most forcibly on the work of the Department of Agriculture relating to the investigation of diseases in live stock.

REPORT OF DR. AUSTIN PETERS, CHAIRMAN OF THE COMMITTEE ON EDUCATION, AT THE TWENTY-EIGHTH ANNUAL MEETING OF THE UNITED STATES VETERINARY ASSOCIATION, WASHINGTON, SEPTEMBER 15, 1891.

In concluding this report, I believe that the Bureau of Animal Industry should receive a little of our attention. I thought of calling your attention to it a year ago, but my paper then seemed so long that I decided to defer what I had to say until a future occasion, and am now glad that I did so, as it has given me an opportunity to beard the lion in his den, so to speak, which I always prefer to do, if the opportunity permit.

We have connected with the United States Department of Agriculture the Bureau of Animal Industry. Its chief is a veterinarian, and a large number of his assistants are also veterinarians. It is the only department in which the United States government officially recognizes the veterinary profession in a manner that at all appeals to our self-respect, and, as the great veterinary organization of this country, we naturally take much interest in its work and usefulness. We are better able, perhaps, than any one else to criticise its actions and results, being, as we are, especially educated on the subjects with which it has to deal. We have the same right as the rest of the people to commend the action of our servants, or to find fault with the way in which they conduct their work, besides which, by our special training, we are in a position to feel that we have a peculiar right to show our approval, or our disapproval, as the case may be, of the labors of this bureau.

Of the practical work of the Bureau of Animal Industry I shall have little to say. It has almost eradicated contagious pleuro-pneumonia from this country, and in time will undoubtedly succeed in its complete extinction. For this service alone it deserves the thanks of the people, and has repaid many times over every cent that has ever been appropriated by congress for its support, including all it has expended in other directions. These results could have been obtained by any good veterinarian possessed of tact and administrative ability. When we come, however, to a consideration of its scientific investigations, we cannot say a great deal for its efficiency.

If we review as briefly as possible the work done in the scientific investigation of swine diseases by the Bureau of Animal Industry, it will be quite sufficient to demonstrate to us the value of its bacteriological work and the credence to place upon any statements emanating from its officials.

If an exhaustive report were written upon the researches in swine diseases in the United States during the past few years, together with all the controversy that they have brought forth, quite a large volume could be easily filled. A year ago, when I thought of referring to this matter in my report, I should have based what I had to say upon an article by J. Armory Jeffries, M. D., which appeared in the *Journal of Comparative Medicine and Veterinary Archives*, for December, 1890, entitled "Etiology of Two Outbreaks of Diseases Among Hogs," although my report was written before the article appeared in print, I was fully cognizant of its contents, having assisted Dr. Jeffries in his work, and in fact done a portion of it myself. Material which I have since been able to avail myself of only confirms me in the views which I then held, without changing them in any important particular.

The other articles of which I speak, and to which I would refer all interested in the matter, as time will only permit my presenting the conclusions I have drawn from them, are:

"A Contribution to Our Knowledge of the cause of Swine Plague, and its Relation to Connected Bacteriological Operations," by Dr. P. Frosch (*Zeitschrift für Hygiene*, Vol. 9, page 235); editors, Dr. R. Koch and Dr. Flugge.

"Upon Our Knowledge of the American Swine Plague," by Dr. Theobald Smith, chief of the bacteriological laboratory of the Bureau of Animal Industry. (*Zeitschrift für Hygiene*, Vol. 10, No. III, page 480.)

"Reply to the Preceding Work of Dr. Th. Smith, upon 'Our Knowledge of American Swine Plague,'" by Dr. P. Frosch, assistant in the Institute of Hygiene of the University of Berlin. (*Zeitschrift für Hygiene*, Vol. 10, No. III, page 509.)

Also at a meeting of the Scottish Metropolitan Veterinary Medical Society, held in Edinburgh, February 25, 1891, Mr. Thomas Bowhill, M. R. C. V. S., read a paper upon "Swine Fever." Vide *Veterinary Journal*, May, 1891. To sum up:

Jeffries concludes that Billings' "swine plague" and Smith's "hog cholera" germs are identical, and differ from those of the disease he has investigated; and that cultures Smith sent him of his "swine plague" germ are identical with the disease germs that he (Jeffries) has been studying, which produce a septic pneumonia in swine that they can communicate to calves, and very probably to lambs, sheep, and other animals.

In short, the much vaunted "swine plague" is simply a septic disease which is not peculiar to swine by any means. It is caused by one of a large group of bipolar organisms, and capable of producing similar symptoms in such small experiment animals as are susceptible to them. Jeffries concludes by saying: "But while only two germs of this class are known to infest hogs in the United States, there may be others in Europe, *e. g.*, 'Wild seuche.'"

I think that Jeffries' work is particularly accurate and very valuable, and am surprised that it has not attracted a great deal of attention, although it does not seem to have done so.

Dr. Frosch, in his first article, compares the work done by Billings with the work supposed to be Salmon's, and draws the following conclusions:

"1. The bacterium of Salmon's hog cholera and Billings' swine plague are identical.

"2. The same is the cause of the American swine plague, while the proof of an etiological relation of the bacterium of Salmon's swine plague to the first, especially to a second plague of like extent, has not yet been sufficiently demonstrated.

"3. That the bacterium is identical with Selander's Schweine pest bacterium (Selander's Schweine pest being the swine disease of Sweden and Denmark), but different from bacterium of the German Schweineseuche, chicken cholera, rabbit septicæmia, and ferret plague.

"4. The ferret disease is caused by a separated kind of a bacterium and cannot be grouped with the rest."

Dr. Smith's is a reply to Dr. Frosch's first article.

Dr. Frosch's second paper is a reply to Dr. Smith.

Mr. Bowhill's paper announces that he has found in cases of swine fever, in England, a bacterium identical with Billings' swine plague germs and that he has sent specimens to Billings, who confirms his discovery.

Here we have two excellent investigators, one in the United States and one in Germany, confirming the identity of Billings' swine plague germ and Salmon's hog cholera germ, and each one acting independently of the other, while the third finds the same germ as the cause of the English swine fever.

Dr. Billings boldly announces that he found his germ of swine plague in July, 1886, among the first pigs that he examined in Nebraska, which had died of the disease.

Salmon, in his report for 1884, discovered a micrococcus as the cause of what he then called swine plague. In his report the next year he says it is due to an oval, motile bacterium. Later, in some of his replies to his critics, he attributes the discovery of this organism to his assistant, Dr. Th. Smith. Dr. Frosch says: "This circumstance not only readily explains the intrinsic contradiction of the reports for 1884 and 1885, but also seems to have influenced Salmon's further investigations."

In a special report of the Bureau of Animal Industry upon hog cholera, its history, nature, and treatment, issued in 1889, there is a short history of the investigations of swine disease made in the United States, but we do not find any mention of the name of Billings, although he discovered at once the bacterium which the chief of the Bureau of Animal Industry had been searching for for years, and which he probably would not have found for some time if he had not had the help of an assistant whom he was not generous enough to credit with the discovery and so let it pass as his own.

In the report of the Bureau of Animal Industry for 1886, page 20, we find the following statement:

"In view of the results of investigations which have shown the existence of two distinct infectious diseases of swine, perhaps of equal virulence and distribution, a change in the nomenclature becomes necessary in order to avoid any confusion in the future. Since these two diseases have been considered as one in the past and the name swine plague and hog cholera have been applied indiscriminately, we prefer to retain both names with a more restricted meaning, using the name hog cholera for the disease described in the last report as swine plague, which is produced by a motile bacterium, and applying the name swine plague to the other disease, the chief seat of which is the lungs. This change is the more desirable, since recent investigations have shown that the latter disease exists in Germany, where it is called swine plague (Schweine-seuche)."

The following questions propound themselves to us after reading the above:

After speaking of the disease as a swine plague for several years, did the chief of the Bureau of Animal Industry call Billings' swine plague "hog cholera" for the sake of creating confusion? (Thus, while apparently ignoring him, at the same time paying him the greatest possible compliment in the power of one man who seems to admire another.)

If the name "hog cholera" was not used in place of swine plague for the purpose of creating confusion, why was a septic pneumonia of the pig termed "swine plague" unless it was for the purpose of causing still further confusion. When, as we have seen, the disease is not confined to swine, but a little careful study would have shown that the pigs could easily communicate it to other species of animals, Dr.

Frosch pays the methods of bacteriological study pursued in the laboratory of the Bureau of Animal Industry the deservedly high compliment of doubting any "etiological relation of the bacterium of Salmon's 'swine plague' to the pest, especially to a second plague of like extent."

But Jeffries' work removes all doubt upon this matter and we know that the Bureau of Animal Industry has found another disease of swine, which is a septic pneumonia and is not alone confined to swine, and which for some reason or other they choose to term "swine plague." Furthermore, it is not impossible that one animal may be infected with both maladies simultaneously.

The so-called swine plague of the Bureau of Animal Industry is one of those septic diseases due to filth and is seen chiefly where putrifying city swill is fed, and farmers around Boston find that if the swill is boiled and then fed before there is time for putrefactive process to commence again that they are not troubled with it. In this respect it resembles closely the German Schweine-seuche. If this be a true swine plague, make the most of it.

Dr. Smith's article is, as I have said, in reply to Dr. Frosch's first article. In it he attempts to uphold the work done under the auspices of the Bureau of Animal Industry and to throw discredit upon the work done in Nebraska and also to answer the criticisms in Dr. Frosch's first article.

Dr. Frosch's reply to Dr. Smith has its chief interest in his closing sentences. After briefly answering Dr. Smith's remarks and saying that there is no need of his defending Dr. Billings, as he is abundantly able to defend himself, Frosch ends with: "From the present publication of Smith's, however, which could not be seen in reading the reports of the Bureau of Animal Industry, it is evident that Salmon was not the discoverer of either the 'hog cholera' germ or that of the 'swine plague,' so now we know the condition of things in that regard."

Whether Frosch's feelings of admiration for the honesty and generosity of the pseudo-scientist, whose work he supposed he was reviewing when he wrote his first article, were equal to his feelings of pity and contempt of his assistant, who was obliged to give the credit for his hard work to his chief, or lose his official head, and yet serve as a pillar for his doughty chief to hide behind in case of an attack, I leave to your imagination.

You will see that Jeffries, in his paper, gives Smith credit for the work he has done. It has been no secret to me for the last year and a half as to who was actually conducting these investigations in the Bureau of Animal Industry. Having taken the investigation of swine diseases as a fair sample of this bureau's scientific labors, are we to be expected to place any dependence upon the accuracy of the statements emanating from its officers concerning such work, especially when they conflict with the results obtained by men like Paquin and Billings, unless the work of the former is confirmed by experiments conducted by independent and unprejudiced recognized ability?

How can we, as a profession, feel anything but disgraced when we think of the opinions which must be held in Koch's laboratory, the greatest bacteriological laboratory in the world, concerning our Bureau of Animal Industry and its scientific work.

I do not wish any one to think that I have taken up the cudgels in Dr. Billings' behalf. Scientific research is the search after truth, and work that is recognized as good abroad cannot be ignored at home, no matter what the personal feelings of one man may happen to be towards another. No one deplores more than I the personalities that so often pervade the writings of the investigator employed by the state of Nebraska, that have done so much to detract from the dignity of his work, which I believe to be really correct and valuable. On the other hand a lack of honesty and straightforwardness is equally bad or worse, and modern political methods are not to be tolerated in the conducting of scientific researches.

The former style of writing shows what it is on the face of it. The latter often hides a good deal beneath its surface. One is like the rattlesnake, which gives warning when it is about to strike. The other is more dangerous, like the deadly moccasin, which strikes its fangs into its victim without giving any indications of its presence.

If the Bureau of Animal Industry is to be a political organization, why not have its chief simply write the letter of transmissal of his annual report to the secretary of agriculture, and have a few true scientists in its employ to work unhampered, and make their own reports upon the questions that they have been studying upon. This, at least, for the sake of making a more creditable appearance to other civilized nations, if we have no respect for ourselves.

More could easily be added of adverse criticism upon the management of the Bureau of Animal Industry, but enough has been said for the present, and it does not seem advisable to continue this report to too great length.

In conclusion, I wish to heartily express my thanks to my confreres upon this committee for the valuable assistance they have rendered me in obtaining material for this report.

AUSTIN PETERS, M. R. C. V. S., *Chairman.*

I have recently taken pains to read carefully all my various publications, and I cannot find a misquotation in any one of them, and only one misapplication, which was perfectly unintentional, and is on page 230 of my bulletin on swine plague, which I much regret. As to the "man" I desired to have sent here by Mr. Rusk, I supposed the gentleman was deep in his confidence from long acquaintance, and I have given my own reasons for desiring him in my second "open" letter to Mr. Rusk. I am sure I am right in asserting that if preventive inoculation is to be of any benefit to the farmers of the west it must be so simple that they can do it themselves, otherwise, while as a scientific fact it is possible, it would still be an impracticable procedure; but of this later on. As to being afraid of any one's seeing my methods, every one knows that so far as the laboratory is concerned they are nothing more or less than the simple transference of the blood from the heart of a diseased hog to bouillon and the use of the same within four days. All other laboratory methods have been found injurious to the preventive qualities of the virus; and hence, one by one excluded as this fact has been demonstrated. Success depends on the selection of the right kind of an outbreak and at the right time in the outbreak, but mostly on an outbreak just beginning, and the pig most recently attacked. In other words, *a fresh outbreak and a freshly attacked pig is the only source from which to obtain virus, because, practical results have conclusively demonstrated that a pig that has been ill over a week, and the shorter time the better, does not yield a reliably preventive virus.* This has all been expressed in the letters to Mr. Rusk, so what was there to observe save the ordinary methods in use in every laboratory, and the practical results of viruses taken from all kinds of outbreaks, and different stages in outbreaks, which was the method pursued last year, in order to settle the above point beyond further controversy. There would then have been no need of resorting to all sorts of disgracefully underhand methods, as has been done by the Department of Agriculture, in order to discover what was going on in this laboratory during the past year. This is a public institution, and had I refused all desired information to the department, the chancellor of the University has access to all records and books of the work done here, as shown by the results of an investigation of the same by a committee nominated at the chancellor's request at the late meeting of the National Association of Swine Experts, which reported as follows:

At the closing meeting, also, the following was submitted by the special committee selected to investigate Dr. Billings' work in inoculation for hog cholera:

Your committee, to which was referred the matter of the work now being done at the patho-biological laboratory of the State University, desires to make the following report:

In company with the chancellor of the University we visited the laboratory and made an examination of the same, its methods and its work, lasting for several hours. Every facility was offered us in the way of opening all books, records, files of letters, reports, and other data in the office of the laboratory. Others were with us at the time and there was the utmost freedom of inquiry and the fullest possible replies to all that we asked. The office was treated as a public office and the examination was conducted in the most thorough way possible.

As a conclusion we desire to say:

First—We are satisfied that the experiments now being carried on in the laboratory are conducted with unusual skill and ability, and with facilities for such work that can scarcely be found elsewhere in this country. We are of the opinion that the state is warranted in continuing these experiments, and indeed that it ought to give more aid that they may be carried still further and more rapidly. We believe that the results of the experiments thus far made and recorded are of great and practical importance to the farmers of this state and the entire nation. We believe that it is especially true of the experiments made in connection with hog cholera by inoculation, and we confidently anticipate that this system of prevention will be shown to be safe and effective. We believe that it is worthy of careful trial in all infected districts. We confidently assert that the records of the laboratory opened to us have conclusively proved that it has been effective in many instances in Nebraska, and to such an extent that it can no longer be called an experiment, nor can it be seriously questioned.

Second—We carefully compared the statements set forth in the Farmers' Bulletin No. 8, issued by the Department of Agriculture, with the originals of letters and reports in the laboratory, many from the same parties said to be quoted in this bulletin, and we believe that the government has not been well nor wisely informed in this matter. We deem the letters in the bulletin to be so misleading as to make it unsafe to accept many of its conclusions. We do not think it would be wise for farmers to be led by the bulletin to avoid an experiment in inoculation. We believe that parties have thorough forgetfulness in the lapse of time or because of the uncertainty of hearsay testimony, been led to report to the government in a way that is misleading.

In conclusion, your committee would urge on all members of this association from states other than Nebraska the necessity and desirability of securing action in their own states like that taken by Nebraska, in the way of establishing state laboratories in which the disease of animals can be investigated. And we congratulate Nebraska on having a university that is willing to give its attention to such matters as well as to mere teaching of those things that are generally taught in colleges. Respectfully submitted,

H. C. OILAR, Indiana.
ALONZO BAKER, Iowa.
L. HAMILTON, Nebraska.

Again, Mr. Rusk says:

> You will find by investigating the matter that neither during his first nor second engagements in Nebraska has Dr. Billings been attacked by Dr. Salmon; on the contrary, Dr. Salmon has been the subject of continued attacks from Dr. Billings.

That that accusation is false and the secretary of agriculture misled has already been made evident in the demonstration of the facts that the agricultural department, under his predecessor, did try to prevent my first engagement in Nebraska by informing the authorities here that they would "send a better man" to do the work, and in the second place by the letter of June 11, '91, of the chief of the Bureau of Animal Industry to the editor of the *American Swineherd*, which, though an attack on inoculation, was still an attack on me and the work I was expected to do on my return to Nebraska, July 1, 1891. While no one will question the right of the chief of the bureau to express his opinions on inoculation as a preventive of swine plague, no fair-minded person will deny to me the right to stand by my own and the interests which I have been engaged to advance. The "right" in the matter, however, entirely depends on the facts and evidence in any case. Whether they have been used honestly or intentionally misconstrued to create public prejudice. That the latter is the case, so far as the chief of the bureau is concerned, will soon be so completely demonstrated that he who reads cannot help but endorse that statement in the same manner as the committee appointed by the National Association of Expert Swine Judges recently did.

The next remark of Mr. Rusk's is the most amusing of all. He accuses me of "having been driven to the wall" and then posing as a "long-suffering individual who has been quietly and disinterestedly laboring in the cause of the farmer."

To these remarks I can simply refer to the action of Nebraska. I am not one who makes "frantic appeals" to any one on my own account. 'Tis true that I have constantly done the best I could under the circumstances to awaken in the American farmer that degree of individual self-respect by which he should stand up for himself and break the chains of complacent slavery which machine bosses, like the secretary of agriculture, have forged around his pliant neck. I have not done this either as Billings, or as an investigator, nor in the interests of inoculation, but as a man who indignantly sees the spirit of the declaration of independence ignominiously trampled under foot and the people made willing slaves in spite of their constitutional rights because of their ignorance and self-complacent blindness. My course and writings on other subjects than diseases and their causes are sufficient evidence of the truth of my statements as to the real reason I have combated the course of the Department of Agriculture and other methods of the political ring-masters; which course I intend to pursue with even more vigor, whether agreeable to the kings of the machine and their minions or not.

Mr. Rusk next alludes to the test made by his department at Ottawa, Illinois, during the past winter in this way:

> After clamoring so long for an opportunity to demonstrate the success of his inoculation as a preventive of swine plague, it was certainly very ridiculous to find him evading the opportunity which Dr. Salmon offered him for making a comparative test at Ottawa. He wanted an investigation, but not of scientists; here was a chance to demonstrate the value of his method to a committee of intelligent farmers, but instead of accepting it with alacrity he did all he could to delay the test.

I emphatically denounce every word of the above as false and maliciously misleading, but am equally pleased that Mr. Rusk has put himself on record as he has. No more villainous and unscrupulous piece of political work has ever been performed by the department than that at Ottawa, Ill. Let us see what was done.

I did not delay the work. I simply refused all connection with it. The farmers about Ottawa had met and were meeting with very severe losses from swine plague in a most malignant form. They first, if my information is correct, in the order of events, wrote to their own Experiment Station at Champaign, Ill., and received the very curt reply that "nothing could be done for them," which naturally leads to the inquiry "Why not?" "For what has that Experiment Station been organized, if not for just such work?" They next turned to the "great father" of agriculture and asked him to aid them in their difficulty, but he said he was financially bankrupt and could do nothing, as shown by the following letter:

U. S. DEPARTMENT OF AGRICULTURE,
BUREAU OF ANIMAL INDUSTRY,
WASHINGTON, D. C., October 5, 1891.

Mr. Geo. C. Cadwell, Utica, La Salle County, Ill.

SIR: Your favor, addressed to the secretary of agriculture, has been referred to me for answer.

I have to inform you that there are no funds available for the purpose of which which you write. The secretary is very desirous of securing an appropriation for the further investigation of diseases among swine, in the near future.

I have directed that the last reports on hog cholera and swine plague be sent you.

Very respectfully, CH. B. MICHENER,
Acting Chief of Bureau.

That is a "flat-footed" refusal enough, is it not, to cause men in distress to look otherwise for relief? Certainly they could not expect anything from that quarter of any immediate value. Then these farmers met, and having heard that attempts, at least, in the direction which they desired were being energetically pursued here, they applied to us for aid.

Did the Nebraska station shut them off abruptly as their own station had done, or refuse for want of money, when it was also nearly "bankrupt," as the National Agricultural Department did? The whole country knows that it did not. We told them to come to us here, as we had no one to send them, and we would cheerfully instruct and do all we could for them. We did so. They sent a man and we gave him all we could spare of the necessary utensils to work with. No instructor on earth can guarantee success in his pupil. Mr. Cadwell is said to have failed and to have killed the majority of the hogs he inoculated, and it is asserted that the disease spread from the hogs inoculated by Mr. Cadwell to the controls and those inoculated by the government. Were this true, which I for one do not for a moment believe, then the government's "superior" or "better" method failed as ignominiously as Mr. Cadwell's was successful in demonstrating to the government that hogs could be killed, in short order, by the inoculation of a culture of the germ, a fact that they have conclusively, and with due regard to exact methods, failed to demonstrate to the satisfaction of the scientific world, as shown by Frosch, and the fact that every authority has given this station credit for such demonstration. Without claiming any more for it, this much can be claimed, that by the department's admission, *a scientist is not necessary, nor the machinery of a laboratory, to obtain an active culture of swine plague germs.* That is one point gained at any rate!

Before we consider that side of the question further, let us go back and continue this history in a chronological order.

After being instructed and supplied by us with the necessary implements, Mr. Cadwell returned home, and what did he find?

That Mr. Rusk had suddenly found an abundance of money and that there was an agent of the department in his locality who had been so busy that the majority of the farmers who had contributed to send him out here, were now turned against the whole thing, and some of them were not very delicate in their hints that "they had been swindled into it and the whole thing was a fraud," which certainly was not a very encouraging reception for a man who had undertaken the journey more in the interests of his community than himself.

Can any one explain why Mr. Rusk was so active in discovering some hidden treasure in the dark recesses of his department, and why he was so suddenly interested in the welfare of these farmers after telling them that he could not possibly do a thing for them for want of funds?

Honestly, now, reader, was not that sudden interest and energy as strong evidence as any one can ask for that the department knows that there is something of value in the Nebraska inoculation, and that for some reason it feared the result, after refusing to do anything, and at once determined to frustrate the endeavors of the farmers to help themselves at any cost, because Billings would get the credit of it?

Can any other interpretation be placed on this hasty change of mind by Mr. Rusk?

The cowardice of a contemptible political trickster was what was the matter in this instance.

Let us see what were the facts in the case: There was a political factor present which must be captured, as against Billings, utterly regardless of cost or principle. That it was done there is no doubt. That kind of game will not win forever on the political checkerboard. Farmers will not forever continue as brainless dice to be played as they please by the kings of the political shaking box. It so happened that the Farmers' Alliance of that district had been the chief mover in seeking aid to overcome the difficulties under which the farmers were laboring. For Billings to have even one lodge of the Farmers' Alliance converted would have been too dangerous a power; hence, the sudden decision of Mr. Rusk to send an agent of his there to "stem the tide." Not only one agent was sent, but the real chief of the Agricultural Department, the boss wire-puller himself, the chief of the Bureau of Animal Industry, and other slaves were sent out also, so great was the danger.

Truly there was another disease more to be feared than swine plague in the Nebraska inoculations. Just think of the result with the Alliance had we succeeded as we hoped to! Refused by their good old "Uncle Jerry" and saved by that fellow Billings in Nebraska! That must never be! Sure enough it was not! In telling their story to the "innocent farmers" of the west, these people have never been honest enough to mention what they knew well enough, that it was absolutely impossible for me to have done anything in person at Ottawa had I been most desirous to do so. It was probably this fact which also stimulated their wonderfully self-interested activity. In October and November I am under engagement each year to lecture to the students at the Chicago Veterinary College, and the whole course is arranged to accommodate me at that time. I was also under an engagement, which could not be broken, to supervise the "lump jaw" trial on the part of the Whisky Trust v. The Illinois Live Stock Commission, a vastly more important case to the live stock men of the west than to prove a thing already proven. Hence the reason for selecting this time by the government. Mr. Rusk says I "wanted an investigation." He knows better! I simply wanted a man agreeable to me sent here by him, to see what we did and how we did it, and then to be at the disposal of Mr. Rusk to send out over the country and instruct where I could not go, because his Bureau of Wonders had done nothing after over ten years' trial.

Mr. Rusk did not dare to accept that offer, simply because the chief of the Bureau of Animal Industry would not let him.

Well, the chief of the bureau issued some sort of a challenge for a test, at Ottawa, Ill., which I ignored, for reasons which have been given, and one other, which was this: *I as well as knew I would not be allowed to succeed under any circumstances, and that the hogs inoculated would be diseased in one way or another by malicious interference;* hence, I would not have inoculated them if I could, and circumstances were such that I could not possibly have done so if I would.

There is not a citizen of Nebraska, who is interested in this work, that does not support my judgment in refusing to have any personal connection with that first public test which the government made of its own method of preventive inoculation, of which it said in 1883:

> Our investigations have shown that the plague is non-recurrent fever, and that the germs might [Notice they are very cautious not to say "have been"] be cultivated; they have even proved that these germs may be made to lose their virulent qualities and produce a mild affection. [How could that be when they only might be cultivated? Were there anything but deceptive promises made out of whole cloth in these words—which we know to have been the case, the government would have said "can be cultivated and can be made to thus lose their virulent qualities and produce a mild affection."] Surely, we have here sufficient evidence to show that a reliable vaccine might ["might" again!] be easily prepared if we carried our investigations but a little way farther. If we had such a vaccine, if it were furnished in sufficient quantities and of a reliable strength, if it proved safe in the hands of the farmer, would not our problem be solved?

The government was forced to publicly demonstrate, at Ottawa, that its promises of 1883 were like those of politicians generally, "all wind." It fell into its own hole and buried itself. It was determined to make any connection of the Nebraska Station with its test an equal failure, because it knew that if left alone there was the greatest probability of our succeding. It knew that, personally, I could do the trick, and did not intend that I should. I knew enough not to try under such circumstances. I do not put my head into the tiger's jaws when I know the nature of his teeth. The inoculation at Ottawa was a fair test of the methods of the Agricultural Department, not only as to their preventive value in swine plague, but as to their honesty and manliness.

The proposition made was that the work should be done by an agent of the department and by Mr. Cadwell, the whole to be placed in charge of a committee of farmers selected by farmers.

The inoculation was done as agreed and Mr. Cadwell went home to his farm, and staid there, seeing the hogs but twice afterwards I, believe. That matters not, for, except as a farmer personally interested, he had nothing more to do with them.

But what did Mr. Rusk's representative do? Did he simply do his business and inoculate the hogs and turn them over to the committee of farmers and return to Washington? No! He staid there controlling the whole thing, talking against the Nebraska method, and stirring up the committee until its members lost what little self-respecting intelligence they had, if they ever had any. Not only this; this committee did as little as the celebrated scientific commission of gentlemen did in 1889. They did nothing. They left the whole thing in the hands of Mr. Rusk's agents, who kept the press busied with reports issued from or through the department at Washington, as fast as anything occurred, or oftener, and to sum up signed a report written for them by Mr. Rusk's agent, unless my informant is like Mr. Rusk's, entirely untrustworthy. I have said that no matter how strongly I might have desired to personally take part in that experiment, that it would have been impossible, from contracted engagements made long anterior to that time. Suppose I had not, does any fair-minded person think I would have gone with my eyes open into such a jungle of political rottenness as that was? I could not have left my work here and gone there and staid night and day for two months, nor could I have sent any one, as we had no funds here; nor would I have trusted the hogs inoculated by me in such hands, unless surrounded by a guard of German troops. I would not trust Americans and left them in the hands of Mr. Rusk's agents, for the very action of the committee shows that they were not men enough to be trusted. On the other hand, had I consented, it would neither have been dignified or honorable for me to have had anything further to do with the experiment. I should have returned to Nebraska at once. Mr. Rusk's actions in this matter are exactly on a par with his permitting the one-half of that notorious commission's report to be sent out to the press in order to create a public disbelief in myself and work and retaining the other half. It would seem that it should be beneath the dignity of any honorable man, much more of a secretary of a department in our government, to boldly trample under foot every principle of honor or manly courtesy.

It is always understood between gentlemen that, when a matter is left to referees, both parties in the dispute shall religiously leave them alone until they make their report. Mr. Rusk seems absolutely without knowledge of the existence of such generally accepted principles of manliness and honor. That he treats them with contempt, and defies any esteem for such principles in the American people, whose servant he is, is so self-evident in the course permitted by him at Ottawa, Ill., as to be indisputable. The most common 'sporting men' have more manly principle and honor than this boss political ring-master seated in Washington. Truly, the stock in trade of these bosses is, *that the people are all fools.* That that committee of farmers should be condemned as such needs no questioning. They were simply a committee of weaklings hoping to draw nourishment for their half-starved intellects from the public teat, with neither intelligence enough nor dignity enough to be honored with the name of men. Fair samples of the poor sticks of to-day who are daily willing to sell their birthright for a mess of political pottage. A country that has to depend for its advance on such stuff as that had better sink into oblivion. Thankfully there are still men enough born to overcome the malevolent influence of such dead-heads. It is a wonder to me that none of the live stock and agricultural editors of the country have appreciated this terribly despotic assumption of power by Mr. Rusk and the disdainfully high-handed manner in which he permitted his agents to act and was led by the nose by his political boss, the chief of the Bureau of Animal Industry. Really, the latter is the only person in the whole matter worthy an iota of respect. We can respect ability even though displayed in wrong directions. There is no question but what the chief of the Bureau of Animal Industry owns the secretary of agriculture, body and boots, and that he is one of the cleverest wire-pullers and most dangerous enemies to the rights of the people of the whole machine which grinds out the links of slavery in Washington. As I have said, I must respect this person's abilities, though I cannot praise them; nor have I any fear of them, personally.

I wish now to return to Mr. Cadwell's inoculation at Ottawa. It has been said that he followed the instructions he received here. Mr. Cadwell has repeatedly denied this and said that he was forced to go directly contrary to them and could not obtain virus from the mild outbreak that he was instructed to. He says: "I want it understood that the virus I used was not from a mild outbreak as you directed, and I told the committee so." Mr. Cadwell also says that the hogs used were from four to six months old, and that he gave them one-half of a syringe, one-half cubic centimeter, *which was according to instructions.*

I wish now to show that if Mr. Cadwell's inoculation did cause the disease and deaths in the hogs inoculated by him, that in using the amount of virus that he did *he implicitly followed the advice given by the Bureau of Animal Industry itself.* Therefore, as intimated before, *a simple farmer has made a culture and taught the bureau something.* The chief of the bureau said (*Journal Comp. Med.*, Vol. IX, 1888, p. 148):

> The most interesting of these experiments is our attempt to confer immunity by inoculation. We soon found that there was no indication for attenuating the virus [We have never been told what kind of a virus was used at Ottawa by them for this purpose] because *the strongest virus might be introduced hypodermically with impunity in considerable doses.* And as the stronger a virus is [That's absolutely wrong] the higher the degree of immunity it produces, you can see that there is every reason for using the fresh unattenuated cultures.

We are left somewhat in the dark as to what we are to consider as "considerable doses." Later on we may read, however, in the article quoted from, that "We made many experiments and *found that hogs might be safely inoculated with one-fourth to one-half cubic centimeter for the first dose.*"

In "Farmers' Bulletin No. 8," on Inoculation, lately issued "by the authority of the secretary of agriculture," which will soon be noticed in detail, it says:

> In our experiments we found that a dose of one cubic centimeter, *i. e.*, from fifteen to twenty drops, of the strongest cultivated virus *would occasionally kill an animal.* [How many is that in a thousand?] *From one-quarter to one-half this amount, i. e., from four to ten drops, have been given without serious consequences in any case.* (P. 8.)

It is evident, then, that Mr. Cadwell did not exceed the amount said to be safe in "any case" by the Department of Agriculture. If he did not, it is evident, also, that those investigators really do not know or have not yet experienced what they really meant when they said that "the strongest virus might be introduced with impunity in considerable doses"; for any one who has tried to kill hogs with the "strongest virus," fresh and unattenuated, should know that even one cubic centimeter is often a most unreliable quantity when used hypodermically.

I cannot find one single case in all the reports where a single hog has been killed by the subcutaneous inoculation of a culture of the swine plague germ that would be accepted by the scientific world as a clear and absolutely proven experiment. Frosch has called attention to the same fact as quoted previously. Bolton failed to produce satisfactory fatal effects to demonstrate this germ by the subcutaneous injection of *five cubic centimeters* of a bouillon culture. He says: "We have not been able to produce the disease by subcutaneous inoculations of even quite large amounts, *five cubic centimeters* of bouillon cultures of this bacillus." What is a poor fellow to do who tries to follow the advice of the department?

He who follows these bureaucratic "blind leaders of the blind" will surely fall into the ditch of absurd contradiction, if not into a slough of mental despondency, over their peculiar vagaries.

Above, they have told us that doses of "from one-quarter to one-half cubic centimeter, four to ten drops, have been given without serious consequences in any case."

We should be able to believe that statement, coming from the supposed highest authority in the land, and published, as it is, "by the authority of the secretary of agriculture" of the government of that infallible example of ideal perfection in that line, the United States of America.

It is plain, is it not, that doses of "from quarter to one-half cubic centimeters, four to ten drops, have been given without serious consequences in any case"?

What are we to understand by these words: That where the named dose has been given, even to the extreme limit of one-half cubic centimeter, ten drops, no "serious consequences" have [n]ever occurred "in any case," that we can give such a dose of the "strongest unattenuated virus with impunity" in any case?

If we can show that the government investigators ever claim to have killed one single hog with that dose, then we have shaken their responsibility entirely as to reliable and trustworthy witnesses or reporters of the facts, even as regards their own testimony, have we not?

Let us turn to page 12 of that "Farmers' Bulletin No. 8," where we can read that

> Of eight hogs inoculated *with one-half cubic* centimeter each, *one died in six days;* the remainder survived.
> Of sixteen hogs inoculated with a like dose (*one-half cubic centimeter*) of the same culture, *one died* and the rest remained well.

That's a queer acting germ they have in Washington! I have never been able to find any such that would kill only one of sixteen inoculated hogs and leave the others "well."

But that is not the point!

There is no clipping or misquotation in the above. The entire passages have been quoted.

What is the matter? Certainly no sane man, especially one in such high authority as the honorable secretary of agriculture, Mr. Rusk, would tell such a constituency as the agriculturists of America an outright falsehood? Something must be wrong with a brain which almost at the same instant could write that doses of

> From quarter to one-half cubic centimeter have been given without serious consequences in any case;

and then that

> Of eight hogs inoculated with one-half cubic centimeter one died.
> Of sixteen hogs inoculated with a like dose of the same culture, one died.

Further comment on the reliability of such "authority" would seem to be unnecessary. What, then, killed Mr. Cadwell's hogs and caused the spread of the disease to the others?

It is not generally the thing to give voice to a suspicion only, but it is my firm conviction that, as the "committee of farmers" were entirely false to their obligations, and as the hogs were in reality left exclusively at the will of Mr. Rusk's representative, the latter was more true to his "private instructions," and that quite extra "considerable doses of the strongest virus" were introduced into the abdominal cavities of the hogs with impunity at convenient times when the "shades of night" had fallen and all honest people were "fast" asleep.

In justice to the truth, and for instruction to the public, I must say that inoculated hogs cannot be penned up too closely (it is the same with naturally diseased hogs, as any farmer knows), or the inoculation may work seriously when in hogs allowed to run free; no visible effects may be seen, though the same virus be used. This may have happened to Mr. Cadwell's hogs at Ottawa, but such learned experimenters as Mr. Rusk's should have learned this by their many years' experience. By whom, or how, was the disease carried to the other hogs? That is the great mystery. It shows unpardonable neglect in conducting the experiment. Such things have never occurred with us in a single instance. Or, to make the government take its own medicine, used in trying to make out that inoculation killed the Hess and other hogs at Surprise, Neb., in 1888, the hogs of Mr. Cadwell may have had greater susceptibility and have become infected from those inoculated by the government. The thing is as broad as it is long, for they claim, in their attempts to mitigate virus, that "the reduction of virulence was not very great even after prolonged exposure to a high temperature for more than 200 days." (Report 1890, p. 111.)

The circumstantial evidence, the dishonorable assumptions of the duties of the committee, the remaining on the ground of the government representative when he should have left immediately, the publication of reports when it was the committee's duty, and beyond that, the disgraceful character of "Farmers' Bulletin No. 8," all warrant the assumption that a public official who is low enough in the scale of manly honor to do such things, or a secretary of government who would countenance them, one or both, is depraved enough to be guilty of any dastardly act to accomplish a desired purpose, especially where success to an opponent means political death to them.

These are strong words. It must not be forgotten, however, that they are true and founded on facts. It has been a matter of pride with Americans, that no serious scandal has ever sullied the fair name of our national government. It has been respectable if not brilliant. Were any of these persons private citizens, their words or acts would be of no more importance to me than a summer breeze. But here we have years of false and misleading assertions, publications, and acts so dastardly bold that one can scarcely comprehend how such things can be supported by and "published by the authority of the secretary of agriculture" of the United States. A grave question is at stake. He who shirks a duty imposed by a public necessity is a traitor to every principle of manhood. The authority of the government of the United States is great. The authority of truth is greater, however.

Mr. Rusk closes his letter to Senator Paddock, as follows:

If the state of Nebraska wishes to continue this kind of a man in such a conspicuous position, paying him $3,600 per year, and allowing him to spend two-thirds of her Experiment Station funds, I suppose she has the power to do so; but her people cannot expect, and have no right to shift on outside parties the discredit and disgrace which he brings on her fair name.

Respectfully yours, J. M. RUSK.

The people of Nebraska have answered. They have judged and passed the verdict. It now remains for the people of the United States to read the evidence, consider it coolly, and pass the verdict as to who is a "discredit and disgrace" to the country, Secretary Rusk or the writer?

Farmers' Bulletin No. 8,

ON INOCULATION,

CRITICALLY CONSIDERED.

SUPPRESSED, ERRONEOUS, AND INCOMPETENT TESTIMONY IN REGARD TO INOCULATION PRIOR TO THE FALL OF 1891.

In the previous article we have considered Mr. Rusk's letter to Senator Paddock. We come now to the consideration of the letters published in the "Farmers' Bulletin, No. 8," with reference to inoculations done prior to 1891. There are forty-seven letters in all, most of which apply to this period; thirty-six of which are hearsay evidence, which would not be admitted in any court in the world. It is not my purpose to deny failures or even to try to explain them away, but rather to show why they occurred, in order that others may avoid them. It is as well known to the investigators of the Agricultural Department as to every investigator in the world, that failures do not militate against a success in scientific investigation. But suppose inoculation had been uniformly successful from the start, has any one stopped to think how little we should actually know as to inoculation under such circumstances? Such success as that, were it possible, would be most detrimental to the acquisition of knowledge. It would kill out ambition. People should remember *that the records of the great discoveries which have made our race what it is have been recorded in words of bitter disappointment on the grave-stones of buried and forgotten failures.*

The millions on millions of men whose lives have contributed nothing to the advance of human knowledge lie buried and forgotten in unknown graves. Only those whose lives have blessed the race have carved their own immortality in the annals of history with deeds incomparably more imperishable than obelisk or pyramid. We know not even the burial place of many of them, and yet the lives of such men as Aristotle, or Hippocrates, or Plato, or a Democrates will continue immortal so long as the human race continues to advance and memory is an attribute of intelligence. Let it be at once understood that I have nothing to defend or advocate in inoculation so far as I am personally concerned. Every investigator knows that if right the world will do him justice in the end. For that he can afford to wait. In that regard I have nothing to wait for. My work has been sufficiently accepted and appreciated already by those capable of judging as to its value. Were that all, I should not only keep still, but cease to be an investigator in this field. When Nebraska builds a suitable laboratory I hope her people will be willing to let me retire. My own ambition as an investigator will then have been satisfied. But I may not wait too long for that. The great battle for justice now going on in the social world has always had far more charms for me than this work I have been doing, though it has been the best fitting possible for it. The "call" is getting so urgent that I shall not resist it many years longer. It is that same "call" which urges me on in this combat against manifest and malicious dishonesty in the public service of our government. I shall continue the battle absolutely, regardless of public opinion at present, and confident of complete justification in the future.

In point of fact the government is not making its attack on our work on account of our failures, but on the baseless assumption that we are a public curse, and our work a dangerous element to the hog interests of the country. This accusation of danger, the suggestion behind the lines, that the majority of the hogs that have died of disease after inoculation *have been actually killed by that procedure, is unequivocally false, and proven to be so by the very testimony brought forward by the government.*

Cheerfully admitting the failure in all the instances quoted, either in part or total as any one may care to look at it, the reader will please remember that it is not failure to protect, but the killing by inoculation of which I am most accused, and that out of those forty-seven correspondents *only one man who had his hogs inoculated positively asserts it to be his opinion that inoculation killed his hogs*, Mr. Hess, of Surprise, Butler county, Neb. Why did not the other men who inoculated and lost their hogs at the same time also write Mr. Rusk to the same effect? That story is an old worn out chestnut. There is not an intelligent citizen of Nebraska, who knows the facts, that believes inoculation killed one hog or pig at Surprise in 1888. We all know of men who, when they have once made an assertion of that kind, stick to it forever afterward. A bill was put into the legislature of Nebraska that year nominally to pay Hess and others, virtually to act as a political lever to lift me out of Nebraska. The legislature did not dare to consider it. The case was too rotten even for politicians. A consistent but mistaken adherer to an error is far more respectable than a malicious and intentional liar. Take, for instance, the reports from Richardson county, six men report on what they have heard only. I have a record there of eleven bunches of hogs inoculated by Mr. F. L. Lewis, of that county, besides his own, viz., for Messrs Jordan, Wittwer, Harding, Shearod, Holman, Carpenter, Cox, and Hummel, but we do not find one of those names in the list of Mr. Rusk's correspondents from that county. *If either one of those men thought for a moment that inoculation actually killed his hogs, does any one suppose that some one would not have heard of it?*

But no complaint has ever been made on that score. Therefore, the testimony of the correspondents from that county is next to worthless. Some men have reported of inoculation in counties where there never has been a hog inoculated according to my records. For instance, W. J. Davis, of Dawes county, who, however, says: "I would recommend inoculation," and Byron Street, of Phelps county, who says that "about ten per cent died from the inoculation. *After a time a herd was taken into a yard where other hogs had the cholera in the worst form. Part of the inoculated hogs sickened, but none died.*"

What kind of testimony is that? If anything, it is favorable to inoculation. To whom did the hogs belong of which "ten per cent died from inoculation." How many hogs were there that "had the cholera in the worst form"? How many of them died? How many were in that herd of inoculated hogs that were put with them of which "none died"? If such a thing did take place, did any one ever hear of putting a herd of hogs in with a lot of hogs "that had the cholera in the worst form" and none dying? The inoculations at Lyons, Nebraska, were failures from reasons which we did not then know.

Is this a failure because four or five died a few days after inoculation, or is it any proof that inoculation killed the hogs? Mr. Wright's name is not on my list. Mr. Underwood inoculated 120, and as the "*owners of both these herds claim that the animals have since been exposed and no further losses have occurred,*" one can see no reason why they should complain.

John W. Tohman, Danbury, Red Willow county, Nebraska:
Two herds that I know of have been inoculated. They belonged to C. Underwood and P. P. Wright. One herd contained about 100 head, and there were four or five of them died in a few days after they were inoculated. I do not know the number in the other herd. The owners of both of these herds claim that the animals have since been exposed and no further losses have occurred.

Mr. C. M. Branson, of Lincoln, who "knows nothing of his own experience," writes about some work at the state penitentiary, of which he also knows nothing.

The state prison tried inoculation, and an extensive feeder of cattle and hogs tried it. It was done by Dr. Billings. There were two herds—I think about fifty in each herd. Nearly all died in one herd, and I think none in the other. Some who have had hogs inoculated have told me that they were highly pleased with it, and say they would not risk having hogs without inoculation. Dr. Billings has often assured me that it is a wonderful preventive. I know nothing of my own experience.

Mr. Branson was in the laboratory a few days since and I asked him who "the extensive feeder of cattle" was? He said, "Hon. S. W. Burnham," late county treasurer. Mr. Burnham never had a hog inoculated. We tested some inoculated hogs in his sick herd and succeeded.

The straight facts are, that in the summer of 1888 I personally went to the penitentiary and inoculated some fifty little pigs; they were then "penned," and it was positively understood they were to have ear-tags put in the next morning, as it had been promised that it should be done before they were inoculated. This was not done and no one could tell them a few days after from about 100 more of the same kind. Not one was injured. The only singular thing about this was that later in the season swine plague broke out and all of the pig crop that was left corresponded almost exactly to the number that had been inoculated. Nothing can be claimed for that kind of work, but it is worthy of record under the circumstances. No other hogs were ever inoculated at the penitentiary.

Mr. H. C. Stoll writes a long letter of complaint. His hogs were sick when inoculated. Mr. Stoll told me, at the breeders' meeting at Beatrice last February (1892), that if I would "give him $200 he would say nothing about it." He then offered to take $100. The value of such testimony is self-evident. Mr. Stoll had "been seen." These hogs were inoculated in 1890. In February, 1891, Mr. Stoll nominated me for president of the Breeders' Association at its meeting that year.

G. D. Mullihan, Paddock, Holt county, Nebraska, reports:

Near Creighton, *where I formerly lived*, there were some hogs inoculated, and there are various opinions as to its preventing cholera, but the majority are not favorable to it as near as I can learn.

How much does that man know about it? The owner of the only herd that I know of that has been inoculated in that district writes:

6–16–'92. My sows and pigs are all pictures of health, at least, so say some of the best hog raisers in the country. C. CLINE.

Here is a most striking example of the reliability of the "authority of the secretary of agriculture" on inoculation:

Dr. H. N. Hall, Ayr, Adams county, Nebraska:

The last outbreak of hog cholera in this vicinity was in 1889. Two herds were inoculated. One belonged to W. Lowman, of Hastings, Nebraska. The owner says one died while testing it, *and the rest never did well and were hard to fatten.* The other herd contained ten animals, and in this none died from inoculation. The popular opinion on inoculation in this part of the state is not very favorable. We are waiting for a chance to test it more thoroughly.

Edward Creager, Juniata, Adams county, Nebraska:

Inoculation has been practiced to a certain extent. It was tested in four herds that I know of, an average of five in each herd being inoculated. I cannot say positively how many deaths occurred for thirty days, or how many afterward, but most of the deaths occurred before that period had elapsed. I would not recommend inoculation.

J. W. Coulter, Hastings, Adams county, Nebraska:

Inoculation has been practiced in this vicinity, and particularly in one herd of about 300 head; the number in the other herds not known. In the large herd a few died in about twelve or fifteen days after the inoculation; exact number not known. I am a strong believer in inoculation, but I would advise care in its use. All the hogs on the place should be inoculated at one time that have not been previously inoculated. Everything said in regard to this should be taken with a grain of allowance, for in 1883, 1884, and 1885 my neighbors' hogs had the cholera and large numbers of them died, and mine were not affected, although they frequently intermingled. I thought this was because I treated my hogs somewhat differently, and that I had found a preventive for the cholera; but in 1886 my hogs nearly all died.

Let us see how well these men agree: Mr. Hall says "two herds were inoculated"; Mr. Creager says that "it was tested in four herds that *I know of*, an average of five in each herd being inoculated." Mr. Coulter does not know exactly.

The facts: Mr. Lowman inoculated 234, Mr. Robinson 20, Mr. Jones 130, Mr. Price 95, Mr. Underwood 50; something different than Mr. Rusk's correspondents, who "know" so much, had any knowledge of. These were done at about the same time. We see no letters from any of these men. That evidence has been willfully suppressed in this connection, I have simply to say that the government refers to "Pamphlet No. 3," sent out by Frank S. Billings & Co. when I was in Chicago, in its Bulletin No. 8. The testimony there given by Mr. Lowman as to his hogs being injured, or that they "never did well and were hard to fatten," directly contradicts this statement of Mr. Hall, so that I wrote Mr. Lowman, sending him Mr. Hall's testimony in full, as given by the government, and received the following:

HASTINGS, NEB., May 25, 1892.

DEAR SIR: Yours of the 24th at hand. I do not know how to answer your letter better than to refer you to my letters of Jan. 1 and Jan. 22, '90, which I enclose herewith, as taken from your circular ("Pamphlet No. 3"), which I enclose. These letters, having been been written while we were in the business, ought to be good:

"HASTINGS, NEB., Jan. 22, 1890.

"DEAR SIRS: On November 7, 1889, my first experience with inoculation commenced. On that date Mr. Bassett inoculated over 234 head of hogs for me, of different ages. On about the 10th of December we inoculated these hogs again, except seven head of shoats that we put out with other herds to test. So far as heard from, these shoats are doing well except one, which has died, but the man with whom it was placed says he cannot say whether it was cholera that killed it or not. He says there was no bunch or mark on this shoat when he was inoculated, and is of the opinion that it may not have taken on him. Understand, these seven shoats were put up with cholera hog on infected ground, and were only inoculated once. Four were put at one farm and three at another, and they are still there.

"*The balance of my herd are doing well; in fact, the inoculating did not set them back any, and I have never had hogs do better.* There has been a good lot of cholera the past fall in the neighborhood. I

shall keep on inoculating this coming spring as soon as our pigs are old enough. We have about 70 sows to farrow the coming spring.
"Yours truly, W. M. LOWMAN."

"HASTINGS, NEB., March 22, 1890.
"New circulars received, which are good. My hogs are doing well. We are now commencing to get our new crop of pigs; have about 50 nice ones. You will remember last December I put out a few shoats to test. I placed three inoculated shoats at H. D. Rood's farm. They have been been in their yard with his hogs ever since, and are all right. He told me recently that he lost six hogs out of that yard with cholera after our shoats came there.
"Yours truly, W. M. LOWMAN."

It is evident that Mr. Rusk had this testimony at hand as he used other testimony from the same pamphlet. What can one say to a secretary of a department who will with "authority" authorize such deception of the people as indicated in that one case of the inoculations in Adams county, Nebraska, in 1889? A government which gives its authority to the publication of such false and deceptive reports is too dishonest and maliciously culpable to be continued longer unless torn down, changed, or built over again on an honest basis.

In his letter to Senator Paddock Mr. Rusk says that I "quoted from department reports, garbling the extracts and misrepresenting the position of the writers in order to deceive his readers and to convey the impression that the department had been inconsistent in its published reports." While I deny Mr. Rusk's assertion most positively, I think the reader must now be convinced that the remarks of the secretary of agriculture are manifestly self-accusing, and that the reports of his department are not only inconsistent in all directions but equally consistent in doing just what he accused me of. As further proof in that direction, I clip three "garbled" quotations from Bulletin No. 8, which were evidently so "garbled" in order to deceive his (Mr. Rusk's) readers, and to convey the impression "that inoculation had been a most dangerous or unsatisfactory procedure' to the persons quoted:

The following extracts from letters received by Frank S. Billings & Co., and published in pamphlet No. 3 on inoculation, are also of interest in this connection:

Thos. L. Peifer, Lincoln, Ill.:

Out of the 42 head (of which 25 were pigs, and of which 4 of the latter died and 1 of the large hogs, since inoculating), my hogs have done exceedingly well; they appear healthy, but I can scarcely attribute this to inoculation, as there has been no disease in the immediate neighborhood, so that the preponderance of evidence would not prove much yet with me. (P. 22, Bulletin No. 8.)

That which was studiously left out, reads:

yet with me, "*but I mean to test it more thoroughly next spring and hope to send you a weighty testimonial in its favor.*

"Yours, THOS. L. PEIFER."

As the first part of the letter is also omitted, and as it is of great value as testimony in favor of the consistency of the secretary of agriculture, in giving his "authority" not to publish the whole truth, it is appended:

LINCOLN, ILL., March 10, 1890.

DEAR SIR: As you are endeavoring to know the results of your work I will give you the effects of inoculation on my hogs as near as possible, and impartially, for the benefit of science.

First, I would say that I am inclined to think that the virus used on my hogs must not have been just right, for the fact that there was no effect observed on any of the hogs except one, which got an enormous sore right over the shoulders. I could not say that this was due to inoculation. One other of the large hogs died.

As to the pigs, I could not notice any effect physically, but don't think it stunted them any, as the pigs were quite small when inoculated, and are now nearly large enough for market. I would therefore say that if inoculation has effect either way, it promotes the growth of pigs.

No injury was done that man.

H. A. Lee, Kearney, Buffalo county, Nebraska:

About the 20th of October, 1888, I had 154 pigs inoculated as a preventive of hog cholera. Twenty-four of the above number were at my home farm, and the balance, 130 head, were 2 miles distant, at the stock ranch and feed yards.

During the two years previous to this I had lost the larger part of my pigs during the late fall and winter with cholera, and believing the yards to be thoroughly infected with the disease, I concluded to try inoculation as a preventive. No cholera has made its appearance on my farms since.

As to its immediate effects; I will say that the 24 head at the home farm, whose feed was principally corn, were most of them affected, over one-half showing cholera symptoms. Some of them did not get over it for weeks, and one died. The 130 head, up to the time of inoculation, had been kept almost entirely on oats, and the inoculation produced no visible effect on them.

On the 24th of October, 1889, Mr. Bassett inoculated 143 head of pigs; 137 of them were at the cattle ranch and 6 small, runty pigs at the farms. The operation produced no visible effect on the 137, but of the 6 head at the farm 4 died.

On the 8th of December, 1889, I took 4 of the 137 and placed them with the hogs of John Reddy, of Gibbon, whose hogs were dying with the cholera. One took the disease and died; the other three are still at his farm, and the last time I saw them seemed healthy and were doing well. On the 28th of December last, Mr. Bassett came and wished to reinoculate those which he had before inoculated, saying he feared the virus used on the 24th of October had lost its protective principle. About 135 head were reinoculated; over half of them were sensibly affected, ceased growing, and lost flesh, and there are fully 40 head that have not yet recovered from the effects of the last operation. (Bulletin No. 8, p. 21.)

Left out:

While the virus used in the operation in October was not strong enough to protect when put to so severe a test as that at Reddy's farm (where over 200 died), I still believe that those hogs would have been safe from all ordinary exposure on my own farm.

I think more care should be used in the operation to proportion the amount of virus used to the size of the hog. It is the light hogs that are injuriously affected, and the heavier ones never show it, is my experience.

Respectfully yours, H. A. Lee.

Without the passage purposely omitted the reader would suppose that Mr. Lee was seriously complaining about the results in his hogs. I do not even now think that the inoculation produced the effect Mr. Lee assumes it did. It is far more likely that the natural disease got in its work at the same time that the hogs were reinoculated.

The third quotation is as follows:

Chapin, Ill., March 19, 1890.

Frank S. Billings & Co., Chicago, Ill.

Sirs: In answer to your inquiry of the 8th, will say that with me inoculation has not been the success that I hoped it would be. The first lot of seventy-four did fairly well; two died soon after the operation, and one disappeared: do not know whether he died or not. One of that lot died a few days ago; he drooped around a few days with outward symptoms of cholera. The rest seem all right of that lot.

The last lot of twenty-seven I would pronounce a perfect failure. They never seemed to get over the operation. They keep running down until they die. There has more than half of them died, and I think more of them will die yet.

Truly yours, C. S. FRENCH.

My own comments were omitted; they were:

It is self-evident that this second lot were infected at the time they were inoculated, and, as seventy of the first lot were with them, it would seem as if the treatment in this case must have been successful rather than the contrary.

In that same pamphlet was published a large number of very strong letters from farmers who had generally tested their inoculated hogs, and it would seem natural that if those unfavorable were selected out, some of the others should also be published, if the "authority of the secretary of agriculture" extends to telling the whole truth to the farmers of this country. Here are a number of the omitted letters:

Bureau County, Illinois.

On October 17, 1889, Dr. Billlngs inoculated a herd of 130 thoroughbred Poland China hogs for L. H. Matson, of Princeton, Ill., and three boars for Mr. Matson's father. At the same time we bought ten of Mr. Matson's boars from him to show our own faith in the method. Mr. Matson has tested a large number of his own hogs in outbreaks in his vicinity, as well as sold four of our boars and numbers of his own, warranting them against swine plague, with the most satisfactory results, and recently bought our six remaining boars, considering them good property to hold.

December 23, 1889, he wrote us:

"Of my father's hogs which were inoculated, one died after being inoculated, but he says the boys hurt him in catching him, and he did not die of disease; the others are doing well, and have been out nearly three weeks among another man's hogs that have the cholera, and are all right. I have inoculated pigs up to Mr. Sawyer's, whose hogs have the cholera; they have been there two weeks. My hogs did first rate after the inoculation. I am the only man here that has missed the cholera, and it must be the inoculation which saved me."

On March 17, 1890, Mr. Matson writes:

"I do not think inoculation hurts a pig in the least; my hogs have done first rate, and grew right along. I think they ought to be inoculated twice to stand the severe test farmers seem to be giving them. My father says his boars grew right on as if nothing had been done to them. Our hogs have done well ever since they were inoculated, and all we can say is we have every reason to believe in it.

"Yours truly, L. H. MATSON."

"LINCOLN, ILL., January 15, 1890.

"*Dr. Frank S. Billings.*

"DEAR SIR: I am glad to say that I have the utmost confidence in the new discovery of inoculation. On the 23d day of November your agent, Mr. Seiler, inoculated 120 head of hogs for me. Three days after I could see a change among them, they seemed to have a slight fever, and were a little off their feed. In ten days after my hogs were all doing well, and I can say that I never had hogs do any better. I do believe that a hog that has been inoculated cannot have the cholera again. I think that hog inoculation is just as much a preventive against the cholera as vaccination is a preventive against small-pox in the human race. I further believe that nine-tenths of the hogs that die from disease die from cholera, and all of the so-called hog cholera medicines are nothing more than humbug and an imposition on the farmers. In the last two years previous I lost 200 head of hogs from cholera, and spent over forty dollars each year for medicine, but of no avail. Therefore, I can say that since I have tried your new discovery I find grand results, and sincerely believe that it has saved my herd of hogs from cholera.

"Yours truly, H. J. PEIFER."

The Clear Creek Swine Co.

"DENVER, COLO., Feb. 17, 1890.

"*F. S. Billings.*

"DEAR SIR: This is the tenth day, and there is not a sick pig on the place; in fact, they are growing faster than ever before. *I was just thinking it would be a good custom to inoculate them about once a week, if it makes them grow so; will wait a day or two and if they do not show it, will send back for some more virus.*

"I suppose I did the inoculation all right. I pulled the skin up a little and squirted a syringe full under the skin of the inside of the hind leg, up above the knee joint. Was that right?

"Very truly yours, JAMES I. BOYER."

Mr. Boyer reinoculated this bunch on April 1, 1890.

And is still inoculating. Many who attended the breeders' meeting at Beatrice, Neb., February last, will remember how strongly Mr. Boyer spoke of inoculation.

Mr. R. C. Fulton, Taylorville, Ill., writes: "The reports in your issue of March 19, by C. A. Cantine, A. R. Hubbard, and Marion Ryman, on inoculation, remind me that it is surely due Dr. F. S. Billings, also the farming world, that I should make report of my experience with two inoculated boars sent me for the purpose of proving so far as possible that inoculation is a preventive. The pair of hogs were received about the 8th of February, and at once placed in the lot of about one acre, adjoining which were breeding pens for six sows. In the lot and pens named I had lost forty head of hogs, and yet had a few left when the two boars were placed in pens. These hogs have continuously bedded on the same litter and in the same pen where eight head had laid while sick and dying. I have fed them on ear corn only, and that was strewn on the excrement and cleanings from other infected pens, and for drink they have had slough water, which catches the waste of barn lots and pens above named. This I consider a crucial test, and I felt nothing short of that would satisfy even my unprejudiced mind as to the prevention of swine plague by inoculation. Many of my neighbors have looked on incredulously, and I was even laughed at by more verdant ones, but the laugh ceased, incredulity has vanished, and all admit that *something* prevented, and what else but inoculation? For my herds have been reduced or swept away before. In conclusion, the hogs named are doing fine on the same ground where forty died, and we all say honor to the *intrepid* Frank S. Billings."

Experience with Inoculation in Iowa.

Mr. C. A. Cantine, Quimby, Ia., had his pigs inoculated with the virus manufactured by Frank S. Billings & Co. for preventing swine plague, and gave two of the inoculated pigs to a neighbor, Mr. A. R. Hubbard, to test them. Mr. Cantine writes, under date of February 14: "I saw A. R. Hubbard again to-day. He says the shoats he took from my place are all right. He took them January 27, and is cutting the dead ones up and mine are eating them. He thought when he took them he could kill them, but to-day gives it up. I will have him bring them home at the end of thirty days. If they live until that time there will be considerable work in this vicinity." Under date of March 9, three weeks later, Mr. Hubbard writes: "I received of Mr. Cantine on January 27 two inoculated pigs, selected at random from his herd for exposure in my yards to cholera. I placed them in the same yard and next where my sick ones were, with three of mine which were in a very bad condition (one dying about that time). One especially was rotting alive, the flesh dropping off in pieces from its ears and head, the stench of which was fearful. At the end of two weeks I concluded to get the germs into them some way, so I sliced up a dead pig and fed it to them without cooking. In fact I have tried in every way to kill these pigs, but have not even made them sick. My neighbors have watched this experiment very closely, knowing that I had the genuine cholera, and that these pigs were inoculated, and that there could be no fraud practiced. To continue the experiment I have lately put in two well pigs that I received from my neighbors. One has been in two weeks and was sick yesterday. I cannot tell whether it is the cholera or not, at this time, but it has all the symptoms. I have just bought some more and will put in two to-morrow. I intend to make this trial as severe and also as fair as possible. I will say this: If any man will succeed by any method in giving two of Mr. Cantine's hogs the cholera I will give him $10. Mr. Cantine will furnish the pigs free if they die, but if they live they must be returned to him."

Experience with Inoculation.

To Western Resources: Like all progressive farmers who have gotten done with scrub stock of all kinds, I have watched the reports from experimental farms of the different states, and being a citizen and farmer of Nebraska, was particularly interested in the investigations being made by Dr. F. S. Billings in the swine plague known as hog cholera; and to say that I felt elated when I read of the appointment by the government of a commission to harmonize the difference, as I understood it, between Dr. Billings and Dr. Salmon, don't nearly express it.

I was in hopes that their report would settle the difference between the doctors and give us a remedy that would assist in combating the much dreaded disease. The report of the commissioners gave me no satisfaction; in fact, it left me unsatisfied as to any remedy at all.

I had carefully read all the reports that I got possession of, giving an account of Dr. Billings' inoculation to prevent swine plague, and concluded to experiment, as I then thought, at home.

I went into my feed lot, where I had 120 hogs, and drove out the first twenty shoats that I came to. I moved them fifty rods, placed them in my breeding pens, and had Mr. Courtney inoculate them.

I let them remain thirty-four days, fed them light as per instructions of Dr. Billings, yet they gained in weight rapidly. During the time that I had the twenty in quarantine my hogs in the feed lot got the cholera, and every hog that I had not sold (seventy-four head) got very sick. Out of seventy-four I lost thirty-one, or nearly one-half.

Well, contrary to my intentions, just when my hogs in the lot were dying at the rate of three to five a day, my inoculated hogs got with them and remained with them, slept in the same beds, and to tell it square, two of the inoculated got the disease and died—two out of twenty—nothing near the per cent of the uninoculated.

The eighteen inoculated have never stopped growing, and are

doing very much better than the hogs that had the disease, so much so that I am well pleased with the experiment, and will continue to further investigate inoculation, believing that Dr. Billings is on the road to success. J. H. CURYEA.

Greenwood, Neb.

LELAND, ILL., Dec. 31, 1889.

Mr. Frank S. Billings.

SIR: You must excuse me for not answering your inquiry as to pigs you inoculated for me. I expected to see them a little dumpish, but they got along all right and are doing well. My neighbor across the road had to draw his shoats away, as they were coming down with the cholera. I would like to know if you have inoculated any boar pigs for any firm that has any for sale. I want to buy one, and I want him inoculated if I can find one, and oblige, LEWIS BEND.

LELAND, ILL., March 20, 1890.

Mr. Frank S. Billings.

SIR: Yours of the 8th received, inquiring after the health of my hogs. I have not lost any by swine plague. I had three smothered by the hogs piling together and one by thumps; the balance are all right and doing well. I never had a better lot of shoats than I have now, and I do not know that any of them lost a meal by inoculation.

LEWIS BEND.

Another Iowa Experience.

L. W. Fry, Defiance, Ia., writes to *Iowa Homestead:* "Mr. Maxwell, of this place, and myself suffered severe losses from swine plague during the past season, so we concluded to try inoculation on a small scale, and for the purpose selected small pigs that had never been exposed to the cholera. They were thoroughbred Poland China boars. We have experimented with them until we are satisfied that inoculation will inoculate. I do not believe that inoculation will hurt a pig or interfere with its growth in the least. It did not with ours, and they never were off their feed. After this, we shall inoculate our pigs, and warrant them cholera-proof."

A Kansas Man Tells His Story.

To the Kansas City Live Stock Indicator: Last July, when my herd of swine were inoculated for protection against hog cholera, I promised to inform you of the result, which, I am pleased to say, has, in every respect, come up to my most sanguine expectations.

I have been engaged in stock raising and farming since 1868, with varying success, and during all this time have gained a moderate degree of knowledge of the nature, habits, diseases, raising and fattening of the hog, with a good deal of experience in treating him when sick, which, fortunately, very seldom happens when the dread cholera is not about.

In 1885 a most malignant type of swine plague ravaged my yards, also in 1886, but the loss was lighter for want of material; again in 1887, but still less fatal. In 1888 the dread disease was still present, but, contrary to former experience, it confined its attack to old sows. This year it attacked my young pigs in April, and by May 20 five had died, and most of the herd were coughing. I placed them on new ground, bedded slightly, and soon observed improvement, so that by July 5 they were in a fit condition for inoculation, since which time they have been doing well, better than at any time since 1885. The fall pigs I sold in August, and last spring's pigs will be ready for sale by the middle of December. It is needless for me to say that I am greatly pleased with inoculation as a preventive of the great losses to which I have been yearly subjected. I feel protection, safety, and independence in my business, the same as any other legitimate business.

I wish to say that during these years of destruction to my property I tried every kind of medicine recommended as cholera cures, and never found the least benefit. Experts came to cure the sick ones and prevent the well from becoming sick, but they also failed, and I had about concluded to quit the business, when Dr. Billings wrought out the problem of inoculation.

Perhaps some might think it was not hog cholera, but some other disease that staid so long with me and caused such loss. If there are any such, I wish to inform them that I had a veterinary surgeon open and examine the hogs at different times, and the lungs were always found hardened, the air cells filled with putrid matter, the liver ulcerated, the entrails nearly always inflamed. These symptoms were always present, whether the attack was mild or severe; besides, my own judgment should be of some weight after so long experience.

To my doubting neighbors I offered to put five of my inoculated pigs in any diseased pen they wished, and subject them to the severest test, even to eating the carcasses of those that died with cholera, and that offer is still open and has been published in our local papers.

I hope, Mr. Editor, you may find room in your valuable paper for this imperfect report. If it be the means of saving even one farmer from the loss to which I have been subjected, I will think myself amply repaid.

ROBERT HENDERSON.

Junction City, Kan., Nov. 4, 1889.

These are but a small portion of the letters showing favorable results in inoculation. The selections, from many others in the pamphlet issued from Chicago, were all at the disposal of Mr. Rusk. Any one who reads that Bulletin No. 8, and compares it with the evidence here set forth, must at once admit that everything tending to throw discredit on our Nebraska work has been published and the testimony in our favor largely withheld from the people. Do not, for a moment, lose sight of the fact that every particle of testimony, and probably more, that has been here published as favorable to inoculation was in possession of the Agricultural Department and withheld with the "authority of the secretary of agriculture." This we are justified in assuming, because the "garbled" selections against our work were taken from the same source and published with the same authority.

Two Letters From Iowa.

E. J. Currier, Harlan, Shelby county, Iowa:

In the fall of 1889, November or December, F. S. Billings, by his agent, Mr. Courtney, inoculated 133 young hogs for me. The inoculation was repeated about sixty days after. Between one and two months after the hogs began to sicken, and about 70 of them died. I sent 21 to another farm, and after they had been there a month they took the cholera and gave it to the healthy hogs already on the place. There was no other case of cholera in that region, and my neighbors were not losing any at the time mine were sick. Did inoculation do it? It looks like it. At any rate, I shed no tears because Billings has shut off the supply of virus for all outside of Nebraska. (Bulletin No. 8.).

"HARLAN, IOWA, March 28, 1890.

"*F. S. Billings, Chicago, Ill.*

"DEAR SIR: Yours of recent date as to the condition of the hogs Mr. Courtney inoculated for me, came duly to hand. Would say in reply, that two died of the second inoculation; probably, however, it was their first dose, as our second count slightly overran our first. The hogs did not seem to thrive as well for about two weeks, but got all right afterwards. Two of Mrs. Stiner's died, small ones, and others of the same litter lost their hair. She is well satisfied, however, as the cholera, which was on adjoining farms, did not touch her hogs. I am well satisfied myself, so well satisfied that I have not taken any pains to place any of my hogs in infected herds. The satisfaction of a feeling of security is worth a good deal to a nervous fellow like myself.

"Yours truly, E. J. CURRIER."

From late advices we learn that this herd is not standing as well as we could wish, though the loss is small in comparison to a general outbreak. (Pamphlet No. 3.)

Time and "inflooeuce" make a great difference in some persons' opinions.

THE SURPRISE AND OTHER INOCULATIONS IN 1888.

It is singular that the Agricultural Department should still persist in promulgating a lot of falsehoods about those inoculations. The very fact of my recall to Nebraska and indefinite re-engagement, is ample evidence that no intelligent person in Nebraska believes that any hogs were injured by the inoculations which the secretary of agriculture claims killed 45½ per cent. Here is what they say:

Name of Owner.	No. Inoculated.	No. Lost.
Surprise, Neb.:		
D. L. Sylvester	93	73.
Ed. Hinkley	163	No information.*
F. W. Luddon	52	None.
L. E. Luddon	46	Nearly all.
H. H. Hess	260	220.
C. H. Walker	11	None.
Gibbon, Neb.:		
W. A. Rogers	10	
S. C. Bassett	22	
H. A. Lee	154	
Humphrey & Harris	34	
D. P. Ashburn	18	
Lincoln, Neb.:		
State Farm	30	
Falls City, Neb.:		
Mr. Steele	121	Large number.
Total	1,014	

*[We have never known how many Mr. Hinkley lost, as he made no complaint at the time and so the number did not get into the papers. The statistics had no value whatever. It was a question of killed or not by inoculation. B.]

Mr. Walker has since stated that in L. E. Luddon's herd all but 6 died, and that some of Mr. Hinkley's were lost, but the number was not given. An article in the Omaha *Bee* at the time stated that Mr. Steele lost 110 within thirty days. Mr. Hess states that his total loss was 240. This would make the loss from the information at hand 463 out of the 1,014 inoculated, or 45½ per cent. This does not include Mr. Hinkley's loss, which is unknown.

When it is considered that the experimenter had asserted for nearly two years that he could prevent the disease by inoculation, that during this time the question had been contested and he had been perfecting his method, and that these experiments were made to demonstrate the value of the method, such a complete and disastrous failure in the results is certainly surprising. Under such circumstances it is self-evident that more than ordinary care would be observed in preparing the virus, and in having the conditions as favorable as possible for success.

An attempt has been made to explain these losses on the theory that the herds were infected before they were inoculated, and that the inoculation had nothing to do with the production of the disease. It is said that in one herd several had died before the inoculation; that the two Luddon brothers, who were among those that inoculated, lived side by side, the road only separating their door-yards; that their hogs were inoculated at the same time and in every particular alike, using virus out of the same bottle, yet not one out of the larger herd sickened perceptibly, while with the other herd all but six died; that another brother had a dozen or more hogs that were not inoculated and were kept in a tight pen on the premises with the latter herd, to which they were in no way exposed, "but simultaneously with them sickened and died in about the same ratio."

The following are some of the letters published with regard to these cases:

SURPRISE, NEBR., February 2, 1892.

SIR: Yours of the 28th ultimo received, and I will try and give you my experience with inoculation. The fall of 1888, some time in October, as near as I can remember, Dr. Billings, of Lincoln, sent Dr. Thomas here to inoculate my hogs, which numbered 260. In about eight or ten days after they were inoculated they all took sick. Within four weeks 230 died, and between that time and spring 10 more died. Only 20 survived out of 260.

My hogs were perfectly healthy when inoculated.

That, in brief, is my experience. If Dr. Billings had not been endorsed by the state I should never have allowed him to inoculate; but he had stated in a lecture that it was no more an experiment, but a settled fact; that it was a preventive. I do not believe that making the virus out of a cholera hog and putting it into a healthy hog will work.

I could not recommend inoculation.

Very respectfully, H. H. HESS.

Hon. J. M. Rusk.

D. P. Ashburn, Gibbon, Buffalo county, Nebraska:

Inoculation has been practiced in this vicinity by 6 or 8 persons, having from 30 to 200 animals in a herd. With one single exception none were lost. H. A. Lee, of Kearney, lost 3 or 4 head out of a pen of 24 that were closely confined and had only dry corn and water to eat. He also inoculated about 125 that were running after cattle in a field or large corral at the same time and with the same virus, and the effect was not noticeable. None died or were sick. I would recommend inoculation in careful, intelligent hands, but not otherwise. It creates a mild case of cholera, from which the disease will spread if not prevented, and as the average hog-raiser is not to be relied upon in this particular, I think for general use as a preventive it would be likely to create as much loss as it would prevent. I have used it for several succeeding years with success, and if I again raise hogs shall use it if nothing better offers. I am impressed with the great need of a safer virus, and think it possible that scientific research might discover it.

John Reddy, Gibbon, Buffalo county, Nebraska:

In answer to your inquiries I must say none of my hogs were inoculated, but my neighbors put 8 hogs in my yard as a test that were inoculated by S. C. Bassett, the agent of Dr. Billings, of Lincoln, Neb. Seven out of the 8 died of cholera, and the 1 that lived had a slight touch of it, but recovered. A very poor showing, as we all thought, since a greater per cent of my hogs lived that were not inoculated at all.

S. C. Bassett, Gibbon, Buffalo county, Nebraska:

A few hundred hogs were inoculated in this vicinity in the years of 1888 and 1889. A less number were inoculated in 1890 and 1891. According to my recollection, seven herds were inoculated, containing from 20 to 150 head in a herd. In the majority of these herds—five, as I remember—none of the inoculated hogs died within thirty days. In the other two herds, 3 in one herd of 20 inoculated, and 7 in one herd of 150 inoculated, died. These experiments were mostly confined to pigs ranging from six weeks to three months old. Five of these inoculated pigs were placed in a herd suffering from one of the most fatal outbreaks of cholera I have ever known, and 3 of said pigs died. On my own farm I inoculated hogs first in the spring of 1888, and with one exception have inoculated all pigs farrowed on the farm since that date. I have had no hogs die from the effects of inoculation; neither have I had inoculated hogs die with hog cholera. From my observation and experience I am strongly of the opinion that of hogs inoculated by the Billings method as now practiced a large per cent may be prevented from contracting the disease, hog cholera. I am positive that inoculation by this method does not kill, does not stunt, does not injuriously affect the hog. Its effects are hardly perceptible to those who care for hogs.

C. Dean, Gibbon, Buffalo county, Nebraska:

I cannot tell just how many herds or the number of animals in each herd that were inoculated. Inoculated hogs died in all the herds so far as I know, but cannot state the number. It has not proved to be a preventive for hog cholera in our part of the country.

William Welland, Gibbon, Buffalo county, Nebraska:

Inoculation has been practiced in this vicinity in seven or eight herds, from 25 to 200 in a herd, in past years. None in 1891. In one or two instances 3 or 4 head died. In several instances hogs have been taken from inoculated herds and exposed. They stood the test where the inoculation was properly done. I would not recommend inoculation in its present condition. I believe the practice and principles are right and will prevent disease, but the great liability of spreading disease by inoculation in careless hands is too great to make its general use practical. What is needed is virus thât will produce the effect without starting the disease.

Isaac N. Ewalt, Falls City, Richardson county, Nebraska:

Prof. Billings, of the State University, inoculated about one-half of a herd for a man in this neighborhood, and about half of them died within thirty days. This is the only herd inoculated in Richardson county that I am aware of. I would not recommend inoculation, as I have but little faith in it. I saw Mr. Steele the other day. He is the man who owned the herd inoculated here. I asked him his opinion, and if he could recommend inoculation. He said he did not know whether it was a preventive or not, as the disease was in his herd when they were inoculated, and there were as many of them died that were inoculated as of those that were not.

The government says that 45½ per cent were killed or made ill by that inoculation.

The Gibbon hogs were not injured, as shown by the testimony from there, and in the table, nor were the thirty-nine done at the state farm with the same virus within a day of each other. Those thirty-nine hogs and the eighteen of Mr. Ashburn's were tested in January, 1889, by inoculating them with two cubic centimeters of the same virus which killed fifteen out of seventeen well hogs in one cubic centimeter dose, and the other two only escaped with their lives. That swine plague commission knew all about that test, in fact saw a portion of the hogs, but never mentioned it in their report. It is singular that all the letters they publish direct from the men about Surprise is a repetition of that old old story from Mr. Hess, and that they publish no other letters from a direct sufferer which is good enough evidence that not one of the others think that inoculation killed their hogs at that time. The reader need not fail to notice that of all the hogs killed, according to Mr. Rusk, by inoculation, this Mr. Hess is the only owner that has ever made complaint. Singular, is it not? It is known that Mr. F. W. Luddon did write Mr. Rusk, and that that letter must have been suppressed because it was not the kind of testimony wanted.

That the people of Nebraska never believed that story as to the injury of the hogs at Surprise by inoculation, especially those who had the direct responsibility in the matter, is thoroughly attested by the following letter of the president of the board of regents of the State University, which was written to the editor of the *Farmers' Review*- Chicago, in relation to this very matter:

HON. C. H. GERE'S LETTER TO "FARMERS' REVIEW."

The publishers of the *Farmers' Review* have referred to me concerning the experiments of Dr. Billings in Nebraska, in the matter of inoculation as a preventive of hog cholera, and requested me to give such information in regard to the matter as I may possess. The letter written to you from this state, as quoted, is wholly incorrect.

Dr. Frank S. Billings was employed by the University of Nebraska, in the summer of 1886, to investigate swine diseases, as well as those of other domestic animals, and pursued his investigations until his resignation last July. The hog cholera or swine plague was the most serious of the infectious and fatal diseases among domestic animals in Nebraska at that time, as it has always been, and the first investigation that the doctor proceeded with was with reference to its cause and prevention.

It was not long before his microscope had detected the cause. It was an ovoid microbe that had been previously tolerably well described by Dr. Detmers, of Columbus, Ohio, while in the service of the National Agricultural Department. He immediately began "cultures" of the microbe, studied its characteristics and habits, and soon began inoculating with it. He found that he could kill sound pigs in twenty-four hours with a large "dose" of inoculating material. He proceeded patiently to experiment with different quantities of the culture until he found the line of safety. For the last year of the doctor's experiments I know positively that no hog inoculated by himself has died from the effect of the inoculation unless he purposely gave a fatal dose.

Now, as to the alleged dying of "whole herds" in southern Nebraska last fall from inoculation. That is a matter that I have carefully investigated. He was called to New York for a month to procure certain apparatus that had been ordered by the board, and in his absence he authorized an assistant familiar with his methods to use material for inoculation already prepared by him, upon sundry herds in Butler county, where he had already inoculated with complete success. He, however, directed the assistant to beware inoculating herds that were already infected with the plague, or that he had reason to believe had been exposed to the infection. The assistant went out there and inoculated several herds for well known farmers that had sent in a request for inoculation, in which not a single animal was made perceptibly sick by the inoculation. Then a farmer living in the neighborhood applied to have his herd inoculated. The assistant, on examining the history of the farm and of the herd, was of the opinion that the herd was infected, and declined to do so. The farmer insisted, said that he fully understood that if the hogs were infected the inoculation would be of no avail, but said he would take his chances, and he wanted to save his herd if possible. The assistant yielded, and made the inoculation. That herd commenced to die a few days after, and, if I remember rightly, over sixty per cent died. Then the same assistant, yielding to the entreaties of a member of the State Veterinary and Live Stock Commission, went down to Richardson county and inoculated a herd belonging to the brother of that member, where the hogs were already dying daily from the cholera.

Of course they kept on dying. These, with a small herd in Butler county that the assistant had been assured were uninfected, and which doubtless the owner thought were sound, are the only animals among the thousands that have been inoculated in Nebraska for farmers at their request that have died. In all other cases the hogs were not taken off their feed. And in all these cases the inoculated swine have been usually exposed to the cholera after the required thirty days has passed, and have been immune, the losses having been probably less than one per cent. And where a certain number of marked animals have been left without inoculation in these herds, they have inevitably died on exposure to a virulent outbreak.

Now this story has to do only with swine that have been inoculated for farmers on their own places, and where the animals have not been under the eye of the doctor. The experiments previously pursued on the farm of the University Experiment Station, conducted with every precaution against mistakes and misadventures, had absolutely proven that inoculation gives immunity from further danger from the cholera, and that a certain amount of virus administered under the skin of an animal would secure almost absolute immunity, without any injury whatever to the animal. We have to-day as fine a herd of hogs on the farm as you can see anywhere, and they were inoculated when very young. It did not make them perceptibly sick nor stunt them, and they have survived repeated exposures to infection.

Dr. Billings voluntarily resigned his position as investigator of animal diseases, on account of the opposition raised against him by Dr. Salmon, of the Bureau of Animal Industry at Washington, and his agents in this state. The regents and faculty of the University were convinced of the great value of his work, and believed that it was a great injury to Nebraska, that the silly jealousy of the Washington bureau was permitted to prejudice the minds of sundry state officials and the members of the legislature against him to the extent of threatening to starve the University by the withholding of the annual appropriations unless he was dispensed with. These threats coming to his ears, he at once resigned.

The letter you quote states falsely that the legislature investigated the matter of those inoculations I have mentioned, of herds that afterwards died. Such an investigation was demanded by myself, but refused. And now when I tell you that the neighbors of the men who were alleged to have had their hogs killed by the inoculation of Dr. Billings' assistant are constantly applying to have their herds inoculated by the doctor's agents, you may, perhaps, understand that farmers have done some investigation on their own behalf.

But any reasonable man, of course, understands that even with the greatest caution herds are likely to be inoculated that have already been exposed to and infected with the disease. This, especially, because it is only those who know that their neighborhood is infected or their own farm is full of the germs that desire inoculation. Such herds will die. The inoculation is too late, and probably hastens the progress of the disease, if it has any effect whatever.

In after years this would not affect the reputation of the operator or throw discredit on the theory. But you can see for yourself that in the beginning of the experiments it would be easy to excite popular prejudice against inoculation, if a herd already infected went on dying all the same after the inoculation. Very truly yours,

(Signed) C. H. GERE,

President Board of Regents, State University, Nebraska.

Again these false assertions found their complete refutation at the hands of the Nebraska Live Stock Breeders' Association when they made Dr. Billings their president for the ensuing year at their late annual meeting, and again the next year, and insisted on his continued engagment in the face of all these "inconsistent" and malicious misrepresentations made "with the authority of the secretary of agriculture."

I wish here only to call attention to the utter disregard of the truth, or almost total ignorance of the facts, displayed by the government correspondents from about Surprise, which is in conformity with the balance of the testimony from the majority of these hearsay correspondents. On the other hand, the total disregard of the authors of this bulletin for consistency most emphatically emphasizes the justice of my former criticisms which have been so unpleasant to Mr. Rusk.

For instance, in the table given by the government of the inoculations made with the same virus at that time, the name of Mr. F. W. Luddon is given as having lost "none," while J. H. Sleeger says that he suffered "the same results" as Hess, and others whose hogs were sick when inoculated. Talmage says the same, and also that Charles Luddon lost equally, who never had a hog inoculated. Again, Mr. Steele, of Falls City, is said to have lost a "large number," whereas Mr. Ewalt, of Falls City, says:

I saw Mr. Steele the other day. I asked him his opinion, and if he would recommend inoculation. He said, *he did not know whether it was a preventive or not, as the disease was in his herd when inoculated, and as many of them died that were inoculated as of those that were not.*

That bulletin needed editing badly.

If the information gathered by the secretary of agriculture as to the condition of crops and stock is no more consistent and reliable than that published in this bulletin on inoculation, then the quicker the people see that the expense is stopped the better. If these samples are any criterion, the Agricultural Reports are not worth the paper they are printed on, and, so far as the work of the Bureau of Animal Industry is concerned, it is evident they are not.

What can we think of the chief of the Bureau of Animal Industry who is so careless of his reputation, or so wanting in honesty, as to allow such glaring inconsistencies to go out in a public document as are here displayed? What can we think of the honor and honesty of a secretary of agriculture who allows such things to be advertised by his official authority? The whole evidence goes to show that it is not neglect or carelessness, but an absolutely wreckless bravado which openly defies the intelligence of the farmers of the United States, and assumes that all they will see is that the Nebraska work on inoculation is either a failure, or so terribly dangerous that no one dare even look at it from a distance.

THE CRIMINAL IMBECILITY OF THE AGRICULTURAL DEPARTMENT SHOWN UP BY THE HON. CHARLES H. WALKER OF SURPRISE, NEB.

I am in receipt of a recent bulletin from the Bureau of Animal Industry on inoculation for hog cholera, which assumes to give accurate and reliable information on that subject. As a paper emanating from the most dignified position of this great nation, in the interest of agriculture, the greatest interest of the country, it is certainly a remarkable production. The question considered is one that involves the loss of many millions of dollars to the class of citizens to whom it is addressed, and the detection of a disposition to prevent anything but a truthful statement of the facts unfortunately not only creates a distrust but carries with it discredit to this great department of our government, as to the reliability of its work and a lack of faith in the honesty of its purpose. It will be a matter of great regret to every citizen of this country who feels pride in our institutions, but more particularly to those that suffered so severely and have felt such confidence in the efforts put forth by the government in the matter of scientific research. Science requires only truth; because of this it is difficult to anticipate a condition of affairs that will lead one devoted to science to forfeit all rights to that honor by the sacrifice of truth. It matters not what the provocation may be, whether it is pure rivalry or revenge, the man who cannot rise above personal considerations, but is willing to trifle with so great an interest, in the hands of an innocent public, to advance himself, is of too small a caliber for so great a trust. Emanating as it does from this official source, the statements made in this bulletin will be taken without question by those who have no other means of information, whereas the most flagrant disregard for the truth is found in the testimony submitted to establish the deductions therein, and in some instances the positive statement of a party to an expression is suppressed and hearsay statement of a stranger is published when the hearsay statement best suits the theory of the Bureau of Animal Industry. Such trifling with the confidence of the people is a crime against a noble trust, and it is a question whether even the chief of the Bureau of Animal Industry, fortified as he is, behind a cabinet officer, can afford it.

To show the total disregard for truth, I will refer to the statements of two correspondents that I am personally acquainted with. E. T. Pliefke, Gresham, York county, Nebraska, says:

"I do not know of any inoculated herds that have been exposed to hog cholera and have not afterward suffered from the disease. Mr. Samuel F. Weaver of Ulysses, Butler county, Nebraska, had a herd of seventy-nine inoculated. These were exposed two weeks after and all died but thirteen. As far as I have noticed, it avails nothing. One herd of forty-eight was doing well when inoculated, and in a week began to get sick and die."

The following is a note from Mr. Weaver in reply to an inquiry as to facts:

"ULYSSES, NEB., June 13, 1892.

"*C. H. Walker, Esq., Surprise, Nebraska.*

"DEAR SIR: In reply to your inquiry in regard to my experience with inoculation for hog cholera I would say that I have never had a hog inoculated, and any statements that I have is without foundation and untrue. S. F. WEAVER."

I am unable to learn anything of the herd of forty-eight referred to. The number does not correspond with that of any herd that has been inoculated in this vicinity. The statement is untrue. Referring to the correspondence of Lewis E. Talmage, of Surprise, Butler county, Nebraska, he says:

"In reply to yours, will say the last case I know anything about was that of D. L. Sylvester, of Surprise. In the fall of 1890 he had seventy-five inoculated, and then sold them to Mr. C. H. Walker, of Surprise, to ship to Iowa. I do not know the percentage of deaths resulting from the inoculation. Miller Bros., of Surprise, in 1890, had some seventy head inoculated and lost almost the entire herd. They also had a bunch inoculated in 1888 and lost a large percentage. Mr. Christ Schroeder, of Surprise, had 250 inoculated, and he told me he lost nearly the entire herd, and the few that did live were damaged. Mr. H. H. Hess, of Surprise, inoculated in 1888 probably 200 head and lost ninety per cent. Wilber and Charles Luddon, of Surprise, the same fall inoculated with the same results. The number of hogs given in each case is from memory."

The following note from Mr. Sylvester throws a little light upon a question that Mr. Talmage says he does not know about, viz., the percentage of deaths resulting from inoculation:

"SURPRISE, NEB., June 14, 1892.

"*C. H. Walker, Esq.*

DEAR SIR: I take pleasuse in giving you all the information I have relative to the action of inoculation on my hogs in 1890. I had 100 head inoculated in August. A single one only showed symptoms that would attract the attention of a careful observer that he was sick and his illness only continued for a few days. I did not lose a hog. In September I had them tested with a double dose of virus. I could detect no symptoms of sickness from this inoculation. In December I sold them for 3¼ cents a pound when the market for like hogs uninoculated was only 2¼ cents a pound. The price was demanded and paid because we believed them to be proof against cholera.

"D. L. SYLVESTER."

The hogs were put in the pens at Davenport, Iowa, and fed out without a single loss. The following letter from Miller Bros. tells the story of their experience in detail:

"SURPRISE, NEB., June 15, 1892.

"*C. H. Walker, Esq.*

"DEAR SIR: Referring to a report in the bulletin of the Bureau of Animal Industry, that we had a bunch of seventy head of hogs inoculated in 1890 and lost almost the entire herd, and that we also had a bunch inoculated in 1888 and lost a large percentage, I would say that the Bureau of Animal Industry has been wrongly informed. We never had any hogs inoculated until 1890, and only forty-five head at that time. We had two bunches of hogs kept in separate feed lots, but had but one bunch inoculated. It happened that three or four days after this bunch was inoculated that the cholera appeared in the other feed lot and every pig in the bunch died. Later on we lost seven or eight of the inoculated lot. Whether they died from inoculation or from contracting the disease as the others did, we cannot say, but this is true; our place had been a hot-bed for cholera, and the probabilities for catching it were great. Be that as it may, no fair-minded man, knowing the surroundings, would condemn inoculation from the experience we had.

"FRANK D. MILLER, *for Miller Bros.*"

Referring to Chris Schroeder's experience this correspondent says he has 250 inoculated. At the close he says "the number of hogs given in each case is from memory." I notice this statement to show the unreliability of the correspondent's "memory." Mr. Schroeder had simply one litter of ten pigs inoculated, instead of 250. Further comment under this head is unnecessary.

The following is a letter from Wilbur Luddon, which answers the statement that in the fall of 1888 he and Charles Luddon lost 90 per cent of their hogs from inoculation:

"*C. H. Walker, Surprise, Neb.*

"DEAR SIR: In answer to your inquiry as to my loss of hogs from inoculation, I have to say that I had fifty-two inoculated in 1888, but that I lost none of them. Some showed symptoms of being sick, but recovered in a few days. Most of them, however, did not seem to be sick at all. Charles Luddon never had any hogs inoculated.

"Yours, F. W. LUDDON."

Mr. Luddon remarked that he had made a similar statement to the Bureau of Animal Industry in reply to the circular letter it sent to him. (Mr. Luddon's reply is not mentioned in the bulletin.)

As to the statement made with regard to the Hess herd it is true that he had it inoculated. The work was done on the same day and within an hour of the same time that it was done on Mr. Luddon's hogs. It was done in the same manner and from the same virus, by the same veterinarian, and the two farmers were on adjoining farms. Now it is claimed that 90 per cent of the Hess hogs died from inoculation. That they died is not denied. The conundrum propounded to the chief of the Bureau of Animal Industry is, If this poison killed Mr. Hess' hogs, why did not the same poison kill at least some of Mr. Luddon's? It is a serious question in the minds of many whether poison acts in that way, killing entire herds in the one instance, or rather 90 per cent of a herd, and not affect the other to cause a single death. If such a freak in the action of poison can be cited elsewhere it may be that inoculation did kill them, but if such action is unknown, it does seem that before it is accepted by men of science that the fatality was the result of this action, that a single precedent at least should be cited to sustain the conclusion.

CHARLES H. WALKER, *Surprise, Neb.*

The government is pretty hard put and seems to be "between the devil and the deep sea" most of the time as to what to say about that Surprise business. At one time it uses its figures in the endeavor to show that inoculation failed to prevent the disease, and at another to show that it killed "forty-five and one-half per cent" of the 1,014 hogs. It is not honest enough to call direct attention to the health of the Gibbon, state farm, Walker, and F. W. Luddon hogs, especially in the text (most of which were afterwards severely tested by their owners), and to tell the reader that these were not injured, nor to tell him that Mr. Sylvester's were sick at the time and inoculated at his own request, and that Mr. Steele's were also sick by his own admission in their report. They dare not tell the truth, for it would expose their own base lies. They think the people are fools and will not see these things, which, so far as the seeing goes, may be true, but they will yet learn what became of the boy who cried "wolf" once too often; these government lies have been published once too often in this Bulletin No. 8.

Leaving the Ed. Hinkley, Sylvester, and Steele hogs out of consideration, we have inoculated with the same virus, the same dose, and within twenty-four hours of each other, 331 hogs, not one of which showed a sign of illness, which, as Mr. Walker remarks, is a most singular affair. No infectious disease works that way. No experimenter on earth ever inoculated 1,014 animals of a susceptible species with the same dose of the same culture of a given germ and saw it kill forty-five and one-half per cent, and not touch in any way 331 others. The very fact that nearly 3,000 were inoculated last fall, and that from August 14 to January 20 not an owner complained of injury, and then but one who had "been seen," as we call it, either in person or had seen some of these government lies, and then because a few of his hogs did die nearly three months after inoculation believed that it was the cause, all goes to show the contemptible folly of the following reasoning brought forward with Secretary Rusk's authority: A person who has had the reputation of being an investigator since 1878 (but who never investigated to any success), and who does not know the difference between natural exposure to an *inficiens* in the fields and by-ways of life, which is never equal on the same day and to given quantity at the same time, and the experimental introduction of that same *inficiens* in a given dose to every individual present, can only be said to be a fit candidate for some retreat for feeble minded. Mr. Rusk endorses the following:

In considering this explanation, we cannot lose sight of the fact that the virus of contagious disease is exactly the one poison which does act in an unequal and apparently erratic manner. When a roomful of school children are exposed to one of their number affected with measles or scarlet fever, every child does not contract the disease, though all are equally exposed. The children of some families will contract the disease, while those of other families will remain free from it. This is not the result of living on different sides of a street or in different parts of a town, but it is due to the difference in susceptibility, which varies both with individuals and with families.

What has that to do with the same dose at the same time to all individuals at once? I only wish to say here that all investigators have given too much stress to individual resistance and individual dispositions, which do not exist to natural infections, and not enough attention to the far more common sense fact of irregularity of exposure and slowness of infection and the consequent resistance thereby created in the exposed organism in many cases of natural infection. Unless through traumatic inoculation the infection is generally gradual and often at intervals in natural exposure, whereas in artificial inoculation it is one and equal in all individuals. Of the thousands of healthy hogs and pigs the inoculation of which I have directly followed, I have never seen an indication of such a thing as varying or individual predisposition, and do not believe it exists in natural infection where the natural infection is due to racial disposition, and not an individually acquired weakness, which only predisposes to such a disease as phthisis.

The above attempt at a comparison between the exposure in natural infection and intentional inoculation is too ridiculous to notice further; but for the benefit of people who possess healthy brains, I must say that altogether too much *a priori* valuation has been given to a supposed idiosyncratic variation in the susceptibility of individuals to natural infection. I dogmatically assert that such a peculiarity as greater susceptibility or great resistance in healthy individuals of a species having a natural disposition to a certain *inficiens* does not exist. The phenomena which lead to such an hypothesis are caused by the irregularity of exposure and what might be called fractional infection, some being sufficiently exposed all at once, while others are infected so gradually as to acquire resistance during the exposure. Racial disposition and acquired predisposition, which is invariably individual, are entirely different etiological factors. Of the thousands of healthy hogs and pigs which have been inoculated under my direction and over which I have watched most carefully, I have never seen such a thing as "individual predisposition" predominating in a single case; mind you, I say healthy animals. I pay no regard to age. One thing more I would insist on, *healthy hogs* running at large. In penned hogs I have seen after effects which did not appear in those running free, inoculated with the same dose of the same virus on the same farm at the same time.

SUPPRESSED LETTERS

FAVORABLE TO OUR INOCULATIONS IN 1891,

NOT PUBLISHED IN

BULLETIN No. 8,

BY THE

AUTHORITY OF THE SECRETARY OF AGRICULTURE.

SUPPRESSED LETTERS.

There has previously been given a communication from C. H. Walker, of Surprise, Neb., which conclusively demonstrates that the Department of Agriculture has not published the true facts in connection with inoculation in swine plague as done under the direction of the Nebraska Experiment Station. In that "Bulletin No. 8" occurs the following passage, which leads one to suppose that all the testimony received by the department has been published. We now wish to show that very much favorable testimony has been suppressed and that the whole story has not been told by any means:

> There are a considerable number of owners of inoculated herds in the list from whom the department has received no replies, and it is therefore probable that full returns would considerably increase the percentage of loss as above given.

Surely from that language we are justified in assuming that the eight letters published from correspondents regarding our work in 1891 was all the Agricultural Department received.

In an open letter to Chancellor Canfield, of July 16, received since this manuscript was written, Mr. Rusk admits receiving other letters as follows:

> There were a large number of letters both for and against inoculation which were omitted because they were written entirely from a theoretical point of view and made no reference to any facts in support of the position taken, or for equally good reasons.

That the writer of Mr. Rusk's letter stated a falsehood will be shown by a large number of letters from men who actually had had practical experience and from others who have watched this work and are in a large measure personally and officially responsible for it. Remember the essential accusation is that inoculation has killed hogs and extended the disease.

Col. F. M. Woods, the popular live stock auctioneer of Nebraska, was written to by the department, and yet we see no mention of his answer, which he writes was in this wise:

Dr. Billings.

DEAR SIR: I have the greatest faith in inoculation and in your ultimate success. It is to just such men as yourself that the world is indebted for all great discoveries, and I wish you all success.

Mr. Ed. Slattery, Douglas, Neb., writes us that he answered two circulars that he received, and that his hogs are doing well, and that he has lost none since he inoculated.

Mr. Ezra Wilter, Grafton, Neb., inoculated 195 pigs and hogs in the fall of 1891. When called upon to report in January, 1892, he wrote to us:

We have only been on the present farm long enough to raise one crop of hogs, but will say that the former owner was always having swine plague; in fact, so badly that he gave up raising hogs.

As to danger or injury to his hogs he says:

Not a particle of either; it is absolute carelessness to injure the pigs where the directions are closely given. I have the same faith in it as in vaccination in my family.

Now let us see on what Mr. Wilter pinned his faith. In a detailed letter to us he says:

We live on four corners; on one corner, within four rods, lives Mr. E.; he lost thirty out of sixty head. Mr. T., distant about eighty rods, lost all his young pigs and some old ones; and all along in all directions my neighbors who did not inoculate lost at about that rate. We have not had a sick pig on the place, and I would not be afraid to drop a diseased hog in my herd. In fact, neighbor E.'s sick hogs were among mine every few days. You will undoubtedly get reports that in some herds where inoculation has been done the hogs died, which will be true, but you have always warned against vaccinating sick or exposed hogs as of no benefit. We inoculated a lot of many such pigs and it did not hurt them a bit.

Now it would seem as if the secretary of agriculture would be delighted to hear from such a man as that, for as Mr. Wilter writes us again in another letter:

My inoculated hogs were fully exposed, many of E.'s sick ones coming among them, and one case in particular was a large sick sow which got in with them, and when we got her out and got her home she died in four days.

But we find no letter from him mentioned in that "Farmers' Bulletin," though Mr. Wilter did report to Mr. Rusk as shown by the following letter:

GRAFTON, NEB., May 28, 1892.

Dr. Billings.

DEAR SIR: I got a letter from Secretary Rusk with stereotyped questions respecting my experience with inoculation. I will say that I thought the questions fair ones, and I wrote to him the absolute facts as I have written them to you. Had I anticipated they were were going to take undue advantage of it I would have reserved a copy, and have no hesitancy whatever to prepare a statement of facts over again and send them to you for publication in any agricultural paper. You must not get discouraged, for every new departure from the old school has always been met with scorn, and not until the man was dead has justice been meted out. Then they piled stones on his grave so heavy that the poor fellow could not turn over.

Mr. Wilter has just inoculated his spring pigs.

Hon. F. I. Foss, of Crete, Neb., inoculated 153 in the fall of 1891 without injury or loss and sold a lot of them for extra prices because they had been inoculated, as the purchaser himself told me. Mr. Foss has just inoculated, June 24, 1892, seventy-five three-months-old pigs. Here is another man whose testimony would have been interesting and valuable to the farmers to whom Secretary Rusk pretended he was telling the truth, according to his oath of office and the responsibilities then assumed. Mr. Foss' name is not mentioned in that "Bulletin." Did he write to Mr. Rusk?

CRETE, NEB., June 4, 1892.

Dr. Billings.

DEAR SIR: In answer to yours will say that on receipt of circular from Mr. Rusk, I answered it fully. Can't remember just what I said, but it was all that was necessary. Anything that I can do to aid your cause please let me know.

Yours respectfully, F. I. FOSS.

Messrs. Matthew and Jenner, of Loup City, Neb., inoculated thirty-one head of all ages in the fall of 1891. They report that two small pigs died from some cause, but do not lay it to inoculation, and about June 15 reported to us: "Have not yet had sufficient experience to give an opinion that would be worth anything, but am satisfied with experiment." They write us that "they have reported fully to Secretary Rusk," and as they say nothing more it must be concluded that they have seen no cause to alter their opinion.

Three herds were inoculated by Mr. Ernest Schoff of Axtell, Neb., for Messrs. Pamblade, Lundren, and Johnson (farmers). He reports to us the loss of thirty out of seventy-three for Pamblade, none out of seventy-seven for Lundren, and forty out of eighty-nine for Johnson, against a loss of 95 per cent in the uninoculated herds in the vicinity. Whether Mr. Pamblade reported on the balance of these to Mr. Rusk we have no means of telling, but he does say, "none of them (his own) died from inoculation." He also says: "Sold the old ones and lost about seventeen out of the fifty-five young ones," which is different than my report, as can be seen.

Mr. A. P. Seymour, of Unadilla, Neb., inoculated forty-eight hogs in 1891 and reported that they were "not hurt at all." He writes as follows:

UNADILLA, NEB., June 6, 1892.

Dr. Billings.

DEAR SIR: In reply to your favor of the 4th inst., let me say that I received some questions from Mr. Rusk, but have not received the bulletin you mention, though I noticed extracts unfavorable to inoculation, in some of the papers. I stated the facts in the case to Mr. Rusk as regards my connection with inoculation. We have not lost a hog from disease to date, though there have been losses within two miles. I shall want some virus in the future to inoculate our spring pigs.

Very respectfully yours, ALFRED P. SEYMOUR.

At Diller, Neb., the following farmers inoculated in 1891:

J. P. Cully, 44—"Does not think it injures hogs at all."

Hon. Wm. H. Diller, 196, ninety-eight two months' old—"None injured; valuable."

J. D. Steiner, 57—"Thinks it valuable."

J. B. Diller, 38—"Valuable and safe."

A. B. Wright, 75—Disease must have been on place at time. Inoculated sows and pigs at same time and the pigs naturally died.

As all this was done at the same time and with the same virus it is logical to suppose that as the 325 of the other owners, except Mr. Wright, were not injured, and as he only lost his young pigs (all died), and as he reports his older ones were made ill, it must be assumed, as the other 325 were not at all injured, that the cholera was in the old ones, and that the double dose received at the time, inoculation being added to the natural disease, caused the death of the young ones. It is fair, however, to let Mr. Wright speak for himself. Here is what he wrote us January 25, 1892:

My hogs at the time of inoculation were extremely healthy and were all doing well. I had about twenty head that would range from 250 to 375 pounds. After the operation they eat but very little for four or five weeks and at the expiration of that time they had not gained any and four or five of them were at least forty pounds lighter and one went nearly to a rack of bones, but all at once that one commenced gaining and fattened very fast. The shoats that were from four to eight months old did not seem to have been hurt any, except that they were sick for a few days. The nineteen pigs, about three or four weeks old, were nice and fat. In about four days the sows were nearly dried up and the pigs began dying, and eventually they all died. Taking the thing as a whole I am somewhat damaged by inoculation, though I have said nothing about it even when asked.

The intelligent reader, especially those who have had experience with inoculation, can judge for himself whether this could possibly have been due to inoculation. To resume: We have 325 pigs and hogs inoculated at the same time and with the same virus in the same place, without a particle of injury. Mr. Wright inoculated seventy-five. According to his letter he had twenty old hogs and nineteen sucking pigs, leaving thirty-six shoats not injured by inoculation and which did not even become sick, and none of the old ones died. This is a singular experience in swine plague. The question is open, did not the inoculation after all save the old hogs and shoats by giving them a power of resistance to the natural disease, which the sucking pigs could not possibly have had to an unlimited dose of both natural and inoculated germs? It is worthy of remark that the virus used in this case did give almost absolute protection and did not injure a hog in nearly every herd in which it was used.

I wrote Mr. Wright about what has been said above, and it did not seem to suit his views. In a second letter, as evidence that the same virus was used as on the others, he says: "J. D. Steiner used it, what was left after inoculating his own, and he did the work according to the printed directions that were sent him." It will be remembered that Mr. Steiner did not injure his own hogs, and considers inoculation very valuable.

Mr. Wright reported to Mr. Rusk as he did to me:

I inoculated seventy-five head of swine last fall, of different ages. The large hogs were damaged. Some of them lost in weight nearly 100 pounds. The shoats from six to eight months old I could see no difference in. I inoculated nineteen sucking pigs, every one of which died. There has been no cholera in the neighborhood since I inoculated. My opinion is that inoculation is of little or no benefit.

The interesting question is, what did the others report, or did they report at all? In this connection I received the following letter from Hon. William H. Diller, who does not say whether he wrote Secretary Rusk for the others or not, but this is what he does say:

DILLER, NEB., June 9, 1892.

Dr. Billings.

DEAR SIR: I wrote some time ago to Secretary Rusk in regard to inoculation and made it as strong as possible in your favor and have heard nothing since.

Yours truly, W. H. DILLER.

Mr. J. P. Cully, Diller, also writes that he reported to Secretary Rusk as he had done to me.

Mr. C. H. Crocker, of Emerald, Neb., also inoculated in 1891, and though his name does not appear in Mr. Rusk's list of correspondents he writes me as follows:

YORK, NEB., June 8, 1892.

Dr. Billings.

DEAR SIR: Yours of the 6th received. In reply will say that I did receive a letter of some description from Uncle Jere, making inquiries. I do not now remember whether I answered it or not; if I did, I stated that my shoats did not show any bad effects from inoculation, that I lost none, and would not hesitate to inoculate again. This is the substance of what I wrote, if I did write at all. Wishing you success in your labor for Nebraska and her people, I am,

Respectfully yours, C. H. CROCKER.

Another interesting correspondent whose name does not appear in the list of Mr. Rusk with regard to our inoculations in 1891 is Mr. John Morgan, of Lincoln, who contributes the following:

LINCOLN, NEB., June 10, 1892.

Dr. Billings.

DEAR SIR: This is what I said to Secretary Rusk: He wanted to know my experience of hog cholera and what I thought of it. Here is what I said: In August 1891, I inoculated seventeen hogs, fourteen pigs ten weeks old and their mothers, three sows. I went contrary to the instructions of the laboratory and they got too heavy a dose of virus, and they got in among the sick hogs that were dying with the cholera and eat them. A near neighbor of mine inoculated five at the same time, and they lived and did well. Later on another near neighbor of mine had fifty hogs and we inoculated them, sick and well together, and we thought it saved about half of them. January, 1892, I inoculated seven pigs of my own and they have lived and done well. If you go according to instructions I think it is a good thing and I believe in it.

Yours with respect, JOHN MORGAN.

I now come to a most interesting and valuable case, the testimony of which has been suppressed in Bulletin No. 8. Mr. A. M. Caldwell, of New Holland, Ill., and a very well known breeder of the best of Poland Chinas, has had more experience, good and bad, with inoculation than any other farmer in the country, with the exception of Mr. Walker, of Surprise, and Mr. Bassett, of Gibbon, Neb. The first time I attempted to prevent swine plague in Mr. Caldwell's hogs was in September, 1889, and failed completely and learned more thereby in that and some others, to be considered in another place, than at any one time in my six years' experience. We tried again and the disease got in ahead of us, but the old saying that the "third time never fails" proved reliable for us, and here is what Mr. Caldwell says of his experience in 1891:

NEW HOLLAND, ILL., January 17, 1892.

Dr. Billings.

DEAR SIR: I have on my farm now about fifty pigs that were inoculated when about three weeks old, and then in about two weeks later they were out with my hogs when the cholera broke out, and they are now on the very ground where most of my pigs died that did die, and all sleeping in the very pen where the sick pigs slept, and I never had pigs do better and look finer in my life. I had at the time I inoculated these three sows that farrowed during the interval between the inoculations and at the time of the second inoculation I thought they were too young, and, as I did not care much about them I let them go. They lived on some six weeks and then commenced to die and I don't think a single one of them is left. My inoculated pigs showed slight symptoms, but only one or two died, and are now, as I stated before, doing exceedingly well. I had bought three breeding boars and three sows since I inoculated my pigs. I have them in pens not infected, and as soon as I am done breeding, which will be soon now, if you think it safe to send virus in such cold weather, I want to inoculate them. I have not bred the sows and will not until after inoculation. I have bred some twenty-five or thirty sows for spring litters and shall of course want to inoculate as soon as three or four weeks old.

In the years 1888–90, etc., my losses were enough to buy a pretty good farm. My hogs were not injured or stunted by the inoculation, and my little pigs also were not injured.

I do not think inoculation at all dangerous, and I should scarcely think of raising hogs on my farm without it.

Yours very truly, A. M. CALDWELL.

Unless the mails went astray Secretary Rusk did hear from Mr. Caldwell, and here is what the latter says about it:

NEW HOLLAND, ILL., June 1, 1892.

Dr. Billings.

MY DEAR SIR: In answer to yours of the 2d I will say I did not take a copy of the letter I wrote Secretary Rusk on my experience and opinion of inoculation as a preventation for hog cholera, but it was in substance this: That I had been inoculating since the fall of 1889, that some of my experience had been just as satisfactory as could be wished and that some had been failures. I then gave somewhat in detail my last year's experience, as it was given to you and published in the Omaha *Bee* and *Farmers' Review.* I said further that on two occasians I had cholera break out a few days after inoculation, but that in neither case did I believe it was due to inoculation, for on each occasion I had a bunch of some fifty or sixty head that were inoculated from the same bottle at the same time, which done exceedingly well and never showed any symptoms of disease.

I said as to my opinion that I was very favorably impressed with, and should continue to inoculate so long as I could get the virus; that I had sent to you for virus to inoculate my spring crop of pigs. I further said that I concluded from my experience that when the virus was right it was just as sure to prevent as that I was living, but that I did not believe it was always possible to tell when all things were right, but that I was convinced that it was a step in the right direction and I hoped it would be followed up till perfected. I added that this was my experience, plain and simple, without a disposition to bolster up any one, and that with the unfortunate quarrel, which seemed to exist between Dr. Billings and Salmon, I had nothing to do, but that I was moved to say what I did from the fact that I had lost heavily by this most terrible scourge and was anxious that some means of relief might be perfected.

I remain yours very truly, A. M. CALDWELL.

Mr. Spencer Day, of North Bend, Neb., inoculated seventy head in 1891, and though his name does not appear among Mr. Rusk's correspondents, he writes as follows:

NORTH BEND, NEB., May 31, 1892.

Dr. Billings.

MY DEAR SIR: Your note received last evening. I did receive a communication from Secretary Rusk, and I see from the envelope under date March 29, the inquiries were pointed and very minute, desiring the entire history of my experience with inoculation as a remedy or preventive of hog cholera. He got, I think, all and more than he wanted. I gave him in detail our experience, and said as a farmer I had perfect and entire confidence in the principle and regarded it as solving the hog cholera question. I did not stop with this. I gave him my experience with Jackson's so-called cure, which I regard as misnamed, as, in my case, it killed all the hogs to which it was given, fifty-seven, old and young.

I saw Bulletin No. 8, but did not find any allusion to my letter, but was not in anywise disappointed, as I was aware my letter was not the kind wanted.

Now, my dear sir, if at any time I can be of any service to you, command me. Yours truly, SPENCER DAY.

Dr. H. G. Leisering, a very skillful and highly educated physician of Wayne, Neb., inoculated fifty hogs last fall, and reported in January that they were not injured in any way and that he "considered it scientific and correct if used as directed in all respects." Although his name also fails to appear in Secretary Rusk's list of correspondents, he writes:

WAYNE, NEB., May 30, 1892.

Dr. Billings.

DEAR SIR: I received a letter from Secretary Rusk and in answer told him that after using the virus I had lost no hogs; that whether the result was due to the inoculation or not I could not say from my limited experience, but that, judging from my knowledge of the disease generally, I considered inoculation with the proper virus the only scientific and proper method of preventing the plague.

If any of my hogs get sick I will order my tenant to try the plan you suggest. It looks reasonable to me and I will give it a trial.

Yours truly, H. G. LEISERING.

Mr. John Campbell, of Nebraska City, Otoe county, Neb., one of the few counties that was heavily swept by swine plague in 1891, inoculated eighty-five hogs in January last. He reported to us: "The

11

disease has been very bad in this section of the county for the past sixteen months." He says: "I inoculated sixty-five shoats. About ten days afterwards they commenced dying. They may have been diseased at the time they were inoculated, as my neighbor's hogs across the road were dying at the time. I think it worth a further trial."

The hogs at Diller were all inoculated with the same virus; not including Mr. Wright's, there were 325. Aside from these we have:

Isaac Pollard, of Nehawka	279
P. J. Donlan, Lincoln	100
W. W. Abbott, Lincoln	79
Matthew & Jenner, Loup City	31
F. I. Foss, Crete	153
Dr. H. G. Leisering, Wayne	50
	692

All inoculated with the same virus without injury. Mr. Campbell reports to Mr. Rusk as follows:

On September 18, 1891, I inoculated sixty-three shoats about five months old, and twenty-two old hogs, with virus received from Dr. Billings, of Lincoln, Neb. In about ten or twelve days I lost one shoat, and a good many of the shoats were sick. In about three or four weeks I had lost fifty-nine of the shoats, leaving me four. The shoats and old hogs had been in separate lots, not adjoining, but after inoculating, which was all done at the same time, I moved all the old hogs further away from the others, with the exception of one lame one, and it died in about thirty days after inoculation, but the other old hogs have never had the cholera. I have not much faith in inoculation.

Mr. Campbell also inoculated twenty-two old hogs, and if he followed the directions, gave them one cubic centimeter. They were not injured and escaped illness. As they were in a different lot, we will not claim inoculation saved them, still it all goes to show that inoculation did not cause the disease in the shoats. The government absolutely ignores this side of the story.

As more evidence showing that testimony in favor of inoculation has been purposely withheld from the public by Secretary Rusk, the following letter from the celebrated Shorthorn breeder, Judge Eaton, of Nebraska City, in Bulletin No. 8, is published, as it bears on Mr. Campbell's case:

I only know of one herd that was inoculated, and they were evidently infected before the operation. This herd belonged to John Campbell, of Nebraska City. His neighbor, Simeon Patton, has a hog yard just across the road, four rods distant, both being mostly in a low swale. Mr. Patton's hogs had cholera and were dying fast when Mr. Campbell got the virus and inoculated his own hogs. Mr. Campbell inoculated twenty large hogs and sixty-five pigs or young hogs. Of the large hogs nineteen were kept at some distance from the others and from Mr. Patton's. None of these showed any indications of being sick. The other large hog, being lame, was kept with the pigs. He died, and so did sixty or sixty-one of the shoats out of the sixty-five inoculated. The shoats got sick in six or seven days after treatment and died soon afterward. Mr. Campbell does not believe in inoculation.

Mr. Eaton writes:

Dr. Billings.

DEAR SIR: I think I wrote this communication to the department just as it is printed, in reply to questions. I got the statement from Mr. Campbell, and think the facts as stated therein are very favorable to you and your method of inoculation, except the last line, a mere opinion of Mr. Campbell. In all my communications to the department in reply to questions I have always expressed myself in my own way as decidedly favorable to you and stated my great faith in your methods. I have always had a warm word for you either when speaking to your friends or your enemies, especially the latter. The way they came to get the statement about Mr. Campbell's hogs was in reply to the printed questions. I had that bulletin from the department, but have mislaid it. I am prepared to stand by you in all things, except lumpy jaw, but any other kind of jaw you can give us all right, only I would not like to feast on the beef of a lumpy jaw steer. Science and trade and taste are different things. I mean by standing by you that I have faith and knowledge—that your theories and discoveries in regard to hog cholera, corn-stalk disease, Texas fever, etc., are right and will ultimately do much good.

Your friend, in haste, JAMES W. EATON.

As Judge Eaton is one of the regular correspondents of the Department of Agriculture regarding the condition of crops and stock in his section of Nebraska, would it have been anything else than fair to the work of our station to have printed his opinions as expressed, as he says, and not have withheld them? Mr. Rusk should deal fairly and honestly with the farmers in this as well as other matters pertaining to their interest, or else say nothing at all. It is hard to see how a man in his position can possibly afford to pursue the course he has in this matter of our inoculation.

Another regular correspondent of the Department of Agriculture is my esteemed friend, our incomparable secretary of the State Board of Agriculture, ex-Governor Hon. Robert W. Furnas. In December, 1891, Mr. Furnas had occasion to write a report to the secretary of agriculture, in which he said there was more hog disease in the state than ever before, which was a serious error, as we had only three or four counties seriously affected last year and were freer from disease than we have ever been since my advent in the state in 1886. Mr. Furnas' remarks were so twisted by the department as to make it appear that the disease reported by him was due to inoculation, whereas there has never been a hog inoculated in Mr. Furnas' county. These remarks by the government were taken up by the London *Live Stock Journal*, as follows, which can but result in deterring immigrants coming to Nebraska, and thus injure the state:

London, January 8, 1892.—Mr. D. E. Salmon, chief of the American Bureau of Animal Industry, has issued a report on some experiments carried out in Illinois for the cure of hog cholera by inoculation. Some of the pigs were inoculated by a method recommended by Mr. Billings, others by that of the bureau, and a third division of the same lot were not treated. Three of the first division died in a short time, while the pigs not inoculated remained healthy. But as the two inoculated groups were kept together from the first, and the three pigs were infected by inoculation before the protective effect of inoculation, if any, had had time to mature, the experiment was a failure. Mr. Salmon, however, says: "The results already obtained demonstrate the danger of spreading the disease by inoculation, and particularly by the method used and recommended by Mr. Billings. This danger has been indicated by other inoculations made in Nebraska and Illinois, but it has never before been so clearly and incontestably proved. In Nebraska, it appears, where inoculation has been extensively practiced, the losses of pigs from disease was greater in November than it has ever been before, according to ex-Governor Robert W. Furnas, statistical agent of the Department of Agriculture for Nebraska."

Knowing that Mr. Furnas had always been one of my most persistently strong friends, even when the black clouds of public disapproval caused me to leave in 1889, I had some curiosity to know what he had written the department, and so wrote him, and received the following reply:

BROWNVILLE, NEB., Jan. 30, 1892.

MY DEAR DOCTOR: Reply to yours of the 26th. While a great admirer of you personally, and having great possible faith in your scientific attainments and investigations, particularly in the matter of hog cholera, you must excuse me from any controversy. I have already written Secretary Rusk, in substance, what you now wish me to write you, with the exception of saying I was mistaken. I distinctly stated in my original letter to the department that the disease referred to "was not hog cholera." That I reiterated in my second letter to Mr. Rusk, and at the same time stated there had been no inoculation in my county, and that I had the utmost confidence in inoculation when done by you, and in the principle involved. Now, doctor, what more can I say?

Doctor, allow me, as your friend, to repeat what I said to you in person when last in Lincoln; you are paying entirely too much attention to Salmon.

Again, doctor, I say, go slow. You are all right, and the people who know you are right, and are with you.

Very truly, ROBERT W. FURNAS.

Mr. W. C. Dietericks, of Rockville, Neb., inoculated forty-four hogs in 1891, and in January, 1892, Mr. Dietericks reported to the question, "Were any of your hogs injured or stunted by inoculation?"

Answer: "Can't say that they were; they look thrifty and healthy. It did not have any noticeable effect. I have lost two little little pigs since inoculation. I deem it a valuable procedure on my farm."

Later on he reported to Mr. Rusk:

I had about forty head of shoats inoculated last fall with virus and instruments sent to me by Dr. Billings, of Lincoln, this state. Two little pigs died soon after being inoculated. Do not know if inoculation was the cause. None died of the cholera except one, and that one got amongst a neighbor's hogs and staid several days amongst them. These hogs of my neighbor's had the cholera very bad at the time, although they had been inoculated on the same day mine were. *My hogs did not thrive well after being inoculated, and always looked rough and not thrifty, although they had plenty to eat and were running at large.* I do not think now that inoculation is a preventive of hog cholera. Perhaps the virus had something to do with it. My neighbor and I inoculated the same day. The virus I used was in another bottle than his. He lost five or six hogs, and I lost one, but could not positively say it died of cholera as it died at the neighbor's. I do not think I shall want to inoculate again for awhile.

Now here was a direct contradiction, for in his report to me Mr.

Dietericks said, "They look thrifty and healthy." So then I wrote Mr. Dietericks, and strange to say got the following on that very point:

June 3, 1892.

As you know, my hogs were inoculated last fall. I thought when I answered you (January, 1892, inoculated November 1, 1891) *that my hogs looked healthy and thrifty, in spite of my neighbors saying that my hogs did not look as they ought to. I would not say that my hogs were injured by inoculation. At the time that I answered you my hogs were really O K or I would not have answered you as I did.*

Some time after this, and it must have been after his letter to Mr. Rusk, Mr. D.'s hogs did come down with the cholera, showing the virus was useless or at least not fully up to what it is when right, and he continues:

I really don't know how long after that they became sick and about twenty of those that I inoculated died, besides two old sows that I had not inoculated, and their twenty pigs. There, you can see for yourself that your instruments and views don't amount to anything. My hogs were not sick when I received the question sheet in regard to inoculation from Washington, or I would have answered what I write to you to-day. When you find out something that will really prevent cholera, I will try you again. Until then I am

Yours truly, W. C. DIETERICKS, *Rockville, Neb.*

Now only a word as to the cause of this failure. It is impossible for me to follow out my own conditions for the selection of hogs and outbreaks from which to select virus, unless the disease is pending to a great extent around Lincoln, viz, a fresh outbreak of not over a week old and a fresh hog in the outbreak, and I have to depend on farmers to send me hogs and do the best I can with the material sent. Though I can tell very closely how long the hog sent me has been ill, I cannot always know how long an outbreak has existed, and a hog from an outbreak of over a week or two's duration is never reliable for preventive virus. At this time I was away on my vacation, and the work was left to my assistant, who did the best he could, but was not pathologist enough to know how long a hog has been ill by its lesions. I find on reading his notes the hog used at this time presented chronic lesions, and also on inquiry that the outbreak had been going on three or four weeks when the hog was sent. It is by our failures that we learn when we compare their history with those cases in which we succeed. This has been a long and critical task, which is one reason that no detailed report has been made on inoculation since I left Nebraska. I simply was studying and comparing data.

Another man who inoculated in 1891, but whose hogs are known to have been sick at the time, was Hugh McLaughlin, Lincoln, Neb., who is said to have reported to Mr. Rusk :

I inoculated about fifty hogs last fall, of which twenty died after inoculation. The others lived, and did well. They were all together at the time. I have not seen any sick since.

Now we do not know whether that was all that Mr. McLaughlin wrote Mr. Rusk, but the above does not tell the story as reported here. He writes us : "My losses have been about fifteen, not knowing that they were diseased at the time when inoculated, or not supposing they were, but I have about twenty-five left and doing well and I think inoculation a valuable remedy as any more of mine have not died since, and if that's what saved them I think it would be a good thing for every farmer to try inoculation who has had swine plague and it is unsafe to raise hogs any more on his farm."

The following two farmers inoculated their hogs at the same time and with the same virus as used by Mr. Dietericks. They did not stand, for the same reasons, but I must think that when the full testimony is considered that any experienced hog-raiser will admit that inoculation could not have injured them and that it did give some protection even though the virus was not what I wish it had been. The fault I find is that Secretary Rusk in his bulletin does not try to show that it was a failure as much as he does that it was inoculation which killed the hogs, an absolute impossible thing when the time between inoculation and the time of illness is taken into account:

W. Rotton, Unadilla, Neb.:

I had fifty-one hogs inoculated. I have not lost any with cholera. I have not much faith in it, as I sold eight shoats to a neighbor that had cholera some time before, and he lost two with the disease. They all took it, but all the others got over it and are doing well. (Bulletin No. 8.)

Henry A. Dan, Boelus, Neb.:

I do not think inoculation is a preventive. I inoculated on the 1st day of November forty-nine hogs, and from that time on my hogs have not done well, and the latter part of January the cholera broke out in my herd, and I have lost ten up to this date, April 13th.

They have all been sick, and the pigs that were born came dead or, if alive, they did not live twenty-four hours. (Bulletin No. 8.)

We will let Mr. Rotton's story speak for itself. "Two out of eight" is not much of a loss in swine plague, and we have admitted that the virus was not such as to give any absolute degree of protection. The assertion we are combating is that "inoculation has killed a very large percentage of the hogs treated," which is the utterly groundless impression Mr. Rusk has sought to place in the farmers' minds.

Henry Dan inoculated forty-nine hogs on November 1, 1891. "From that time on," above he says, "they have not done well."

About January 20th, 1892, he answered the following questions:

"Were any of your hogs sick, injured, or stunted with inoculation?" "No."

"How did it affect your little pigs?" "Good."

As to his opinion as to its value he answered then: "I don't know yet."

Now, that contradicts his story to Mr. Rusk, that his "hogs have not done well," so far as the inoculation was concerned, up to the "latter part of January, 1892." From then up to April 13, 1892, he lost ten out of forty-nine. All we can say to that is that either Mr. Dan had an unusually light outbreak in his herd or that inoculation did have quite a degree of protection, though not absolute, for to lose ten hogs in an outbreak of cholera lasting from February 1 to April 13, or seventy-one days, is something unparalleled in the history of swine plague as it naturally occurs.

We find only one other report in the bulletin regarding our inoculation in 1891, which has no value in any way except that it shows the man's hogs were not injured, and that he, like many others, got "stampeded" against inoculation by the cry of its terrible dangers sent broadcast over the land by Mr. Rusk:

S. M. Geyer, Seward, Neb.:

I inoculated thirty head in 1891 with virus prepared by Dr. Billings. It failed to produce any effect at all. I have not lost any since. As to my opinion of inoculation, I think it is more apt to spread the disease than to prevent it. (Bulletin No. 8.)

In this place I will not consider here the balance of the inoculations done in 1891; suffice it to say that while there were some failures there were also many successes, and that the year's work answered our purposes and established many causes of failure and the real cause of success, as well as that farmers can be entrusted to inoculate their own hogs. The most rigid honest investigation cannot find evidence that one single hog or pig was actually killed or injured by inoculation. The one point it has been my purpose to show is that Mr. Rusk has either been most seriously misled, or, as this bulletin has been "published with the authority of the secretary of agriculture," that he has given his consent to withholding valuable letters which must have been received, because they were favorable to inoculation; in other words, that "the truth, the whole truth, and nothing but the truth" has not been given to the farmers of the country in this "Farmers' Bulletin No. 8," on inoculation. Nor can the following statement be anything but false, or at the very best misleading, for by reading it I think any one would be justified in assuming that the eight letters here noted, published by Secretary Rusk with reference to our inoculations in 1891, were all that were received by the department. The large number of letters which are not noted, here published, is sufficient evidence of the willful suppression of testimony "by the authority of the secretary of agriculture," who says:

There are a considerable number of owners of inoculated herds in the list from whom the department has received no replies, and it is therefore probable that full returns would considerably increase the percentage of loss as given above.

THE

FAILURES IN INOCULATION

AND THEIR CAUSES.

FAILURES IN INOCULATION AND THEIR CAUSES.

I have been accused of having published no list or account of the failures which I have met with in my endeavors to prevent swine plague by inoculation, which is true, *but the very accusation is an admittance that our inoculations have been frequently rewarded with success.*

As has been stated elsewhere, *the successes of human advancement are marked on the mile-stones of failure along the roadway of the evolution of our race.*

I have not published my failures for one reason only, and that is, that until now I did not know enough as to their causes to justify their publication. What use is their to humanity in the publication of things one is totally unable to explain? The same is true with regard to our successes, but unfortunately for us in this country, the people demand a report of the mere fact and are very little interested in the causes of success or failure. All they want is that inoculation prevents. They have very little patience with delay in anything that promises them relief from distress. My opponents know full well this condition of things, and, being wily politicians, have done their utmost to augment the popular demand on the one hand and to feed the distrust on the other. Their only desire has been to "drive Billings out," and then they would have had the whole field to themselves, as there is no other special investigator of the infectious diseases of animals in the country, and they could then publish what they pleased, whether consistent or not, whether true or not, without fear of contradiction, as the chances of other states taking up this kind of work are very remote indeed. *It is very expensive and absorbs entirely too much of the Experiment Station funds to conform with the selfish do-nothingism of the majority of the attachés to the experiment stations in the different states.*

It cannot be said, however, that I have been foolish enough to deny having made failures, as is shown by the following quotation from my "Pamphlet No. 3," issued while in Chicago:

FAILURES.

Above I have given testimonials of success only. From a strictly scientific standpoint of view, positive successful results are only taken into consideration, because where failures occur there must have been good and sufficient reasons for them. The evidence given is such as to place the question of preventive inoculation against swine plague beyond the dispute of the most skeptical or the most determined and fanatical enemies of the American hog-raisers. Failures? Yes, of course we have made them, and quite a number also. We are often asked "if inoculation is a dead sure preventive"? Nothing is dead sure but taxes while living, and death which puts an end to them.

Nothing is absolute of human creation. Even the laws of nature are not always inviolable, yet we depend on them as such. It is a law that when we cover a mare with a stallion that a colt shall result. But it does not in every case. There are, however, reasons that cause the failure, which does not militate against the law. It must be remembered that when we began inoculation as a business, much against our will, having been forced into it by the despicable intrigues and opposition instigated by a person in the employ of the government, that we had had no experience in shipping the virus, and did not know what accidents might happen to it during transit. Such things have happened, and in many cases we have found out the cause and can in future prevent it. But one thing we cannot prevent, except by stopping business during the winter months, and that is, the repeated freezing of the virus while in the hands of the express. That this must have been the cause of several failures in Nebraska, western Iowa and Kansas, where it was very cold, is conclusively shown by the fact that none occurred in Illinois where the same virus was used, having been shipped at the same time. In Illinois, instead of cold weather, we had a mud embargo all winter. It should be known that every precaution compatible with the most scientific methods is resorted to in the preparation of this virus.

In Mr. Rusk's "Bulletin No. 8" it is said that "the attempt to feed hogs on glucose refuse and protect them by inoculation was the most disastrous failure of all, only because it was attempted on a larger scale." That the attempt was in general, and from a financial point of view, a total failure is absolutely true, but there were two very large phenomenal successes in it which, could the causes thereof be always repeated, would have led to success had the attempt been continued, and the causes of failure in the balance was not to be sought in inoculation but in the character of the food.

It cannot be said that the Agricultural Department had no knowledge that there had been successes at Davenport, Ia., as well as failures, for it is most positively known that one of Mr. Rusk's special agents visited that feeding station and also as to what he reported to Mr. Rusk and other people at Washington. We do not find his name mentioned, or the fact that he went there, in this Bulletin No. 8, and hence have more evidence of willful suppression of evidence favorable to inoculation "by the authority of the secretary of agriculture."

In Mr. Sylvester's letter to Hon. Charles Walker he mentions a bunch of 100 pigs that were twice inoculated without any perceptible injury, and that they were sold and successfully fed out at Davenport, Ia. The second inoculation consisted of two cubic centimeters of a very virulent culture. This was the "starvation year" in Nebraska, when the crops failed. Mr. Walker himself had over 100 hogs that had been treated in the same way and the same time with those of Mr. Sylvester, and as both these gentlemen had been strong adherents I purchased those hogs, about four months after the last inoculation, and placed them in the pens at Davenport, Ia. *We never lost one of them, and put them all in market in Chicago, well and fat.* We never were able to feed any other inoculated hogs at Davenport on glucose refuse. We would inoculate them on farms and then, when thirty days were over, put them on the glucose feed, and in from fifteen to thirty days they would commence to die in a most terrible manner; but if we left some of the same bunches on the farms, they would remain well and we often tested them successfully in farm outbreaks.

Now all this was known to the secretary of agriculture or his employes, because the agent of Mr. Rusk, who visited Davenport, publicly explained it in many places and also told the results of his investigation on inoculation among Nebraska farmers, telling of failures and successes as well, until "shut off" by the boss of the agricultural machine, Hon. J. M. Rusk; and as this agent and myself were naturally so built as to become very close friends, and as for months he has ceased to write me, it looks very much as if Boss Rusk was very much of a despot and political terror in his small kingdom.

The story is worth relating, as it bears so closely on the point at issue. Personally, I am above animosity towards any of my opponents at Washington. As said before, it is a question of honesty in the public service with me, and I shall fight it out on that line until honesty triumphs, utterly regardless of myself. On the other hand, it's a pure bread and butter question, or one of holding office, with my opponents, and nothing is too contemptibly low for them to stoop to or to be published "by the authority of the secretary of agriculture."

At the meeting of the Improved Stock Breeders' Association of Nebraska, at Beatrice, February, 1891, I met Dr. F. E. Parsons, "special agent of the secretary of agriculture," and one of the most charming men I ever met, one whose innate integrity as a man of honor between men was written on every line of his face. As can be seen, I was immediately taken "by storm" with Dr. Parsons. On introducing himself to me and a group of breeders, the doctor said that "he came out at the special request of Secretary Rusk, and that his mission was to endeavor to show what the secretary was trying to do for the farmers, and that he wanted to be on more intimate touch with them." Dr. Parsons also remarked that "Secretary Rusk enjoined him that he was coming into *my* territory and to be careful not to intrench on my rights, or get into any difficulty; all of which was mistaken. I was at that time a citizen of Illinois, though president of the association, as I was also when elected to that office. I have never yet possessed, or tried to possess, a mortgage on any citizen in Nebraska or his opinions. About the first thing Dr. Parsons said to us, or rather to me, was "Doctor, I have been to Davenport." I naturally asked him what they told him? He said they said "it had not gone all right at all times, but," with emphasis, "*they showed me a fine lot of Nebraska inoculated hogs that they said they could not kill.*" Those were the hogs of Messrs. Sylvester and Walker. They gave the impression to Dr. Parsons that inoculation was a "success" at Davenport and he will bear me out that I told him it had been a failure and explained "why," as I shall soon in this place, and "why" we should not go on with it, even though the Nebraska inoculated hogs had shown that we could. Although Dr. Parsons said that he came to tell the farmers what a good man Jere Rusk was, every one there knows that about all he talked of was inoculation and that every opportunity was given him to find out all he desired with no restrictions or private instructions.

In connection with Mr. Stoll, who was so terribly injured, according to his story in "Bulletin No. 8," it was Hon. J. B. Dinsmore, the president of the Shorthorn Breeders' Association, who mentioned to the meeting that "Dr. Parsons was here and desired to know all he could find out as to inoculation." After various gentlemen had told their story some one suggested that "Mr. Stoll had inoculated some hogs." Mr. Stoll had not "been seen" then by the bureau's agent in Nebraska, but, as Dr. Parsons can bear witness, refused to say a word as to his hogs in any way, and I told the meeting simply that "inoculation had failed there," not mentioning that Mr. Stoll put a lot of infected hogs on to my agent, probably innocently, with the frequently mistaken idea that inoculation was a cure. Dr. Parsons then went to Lincoln to inquire about inoculation, and again came here to the State Fair in September, 1891, still inquiring about inoculation, and there are numerous breeders in Nebraska, Iowa, and other western states who know that one of the doctor's standard expressions was:

I don't care what they think in Washington, all these men I have seen, and they are the most responsible and representative stockmen in Nebraska, tell me that it succeeds when done right, and I know these men are not all liars and I tell them down there [Washington] *that it succeeds and that is all there is about it.*

Dr. Parsons told them in Washington "that it succeeds," "that he saw Nebraska hogs in Davenport, where all the others were sick or miserable wrecks, in which it succeeded," and Dr. Parsons was sent to find this out, but not one word do we find from Dr. Parsons in Bulletin No. 8, and as the doctor's "voice is no longer heard in the land" crying "It succeeds in Nebraska," and as he all at once stopped corresponding with me, the natural conclusion is that Bóss Rusk put the machine-screw at work and that my friend had to choose between his own self-interest and shutting up for the truth.

Verily this is a free and blessed government under which we live! *When the American people once learn that it is the foulest blot on the escutcheon of freedom and themselves the most contemptible slaves on earth, because they can free themselves from their machine masters, there may be hope that justice may once again find opportunity to bloom and thrive in the land.*

To finish up with the Davenport case. The reason that the Nebraska inoculated hogs did not die was that they had entirely recovered from the effects of the inoculation and were entirely free from the germs of the disease. Four months at least had elapsed before they were put in the pens at Davenport. The reasons the others died were:

First—Only thirty days had elapsed and not only were some germs left in them, but their intestines must have been still somewhat congested, though they showed no outward indication of illness.

Second—It transpired that the glucose food caused a change in the chemismus of the blood by which the germs again assumed a malignant character and a virulence almost unheard of in ordinary form infection. Nearly all died. I have said before that we tested this by leaving pigs from the same lots sent to Davenport on the farms, where they remained well and withstood many farm tests among diseased hogs. We were not careful enough in going into this glucose feeding business. Had we inquired of farmers, as I did later, we should have found out that when "cholera strikes a herd of glucose fed hogs it is all up with them, they go like prairie fire. It's all good enough when no disease is about," as a large number of farmers told me. This would not have deterred us at the time, however, as we had tested inoculation sufficiently in other places to know that when done by ourselves for ourselves we could rely on it in distillery feeding, of which later.

In order to see what the glucose did in the hogs, I killed a large number of sick and apparently well ones, and found the intestines of the former almost turned to a solid tube, so thick were they covered with the fine stuff in the slop; the well hogs seemed to be able to handle it. Inoculated hogs could handle it, as the Surprise hogs did, if time enough were allowed to elapse so that all danger of the least possible congestion of the intestines had passed off for some time. We gave up because we could not afford to fight against the intrigues and false statements of the secretary of agriculture and his agents. The country is benefited by the result, so no one else has cause to complain. It demonstrated that glucose slops are not safe to feed hogs on in so conclusive a manner that no one should try it except when driven to their use in an emergency of shortness of other food. In sections where no swine plague exists it is safe, but I do not think it nearly as good a food for hogs as distilling slops.

The primary cause of our failure was ignorance of the nature of the material with which we had to do. Inoculation was never more strongly shown that it could succeed than in the Sylvester and Walker hogs at Davenport, Iowa. As Mr. Sylvester says, we paid them "one cent" more than the market price because we knew we could feed them out from some tests that we have made.

While we spent over fifty thousand dollars finding out about inoculation, and do not complain, as the people will reap the benefits of our experience, we could never see what business it was of the United States secretary of agriculture to do his utmost to make us lose money and prevent our business being a success. I shall speak of that later when I come to sum up J. M. Rusk's malfeasances as secretary of agriculture, and shall show him to be anything and everything but a friend of the farmer or a man of any sense at all as a public official. What Jere Rusk is as a private citizen is none of my business any more than it is his what I do, though he has kept me pretty closely surrounded by his agents and spies.

HOW MISTAKES HAVE BEEN MADE IN INOCULATION.

The chief cause of all the mistakes that have been made in inoculation against swine plague has been ignorance of the many phases of swine plague in relation thereto. I started off on a wrong basis, being entirely misled by the directions which Pasteur and the French school had given to all investigations in search of a preventive virus in non-recurrent diseases. As is well known, the method to which I allude is that of "artificial mitigation." It is doubtless true, or at least appears so, that in diseases which, so far as we know, almost invariably present to us an acute malignant character, or in diseases which are due to spore-bearing germs, like anthrax or rabies, if preventive inoculation is to be possible, artificial mitigation must be resorted to, but there is no reason why this method should be an absolute law—in fact there is every reason why it should not be. In all cases of this kind we are too apt to be blind followers of authority. In seeking preventive inoculation in swine plague I have not made that mistake; in fact my nature is such that I am a very poor follower of any other authority than my own conclusions on the evidence, no matter who or what the authority may be. In swine plague I have never used an artificially mitigated virus prepared with that purpose, though I have very often used a virus that has been robbed either in part or *in toto* of its preventive power because I did not know two things:

First—*That the assurance of possessing a mild, non-pathogenic degree of virulence in a virus was absolutely valueless as a guide with regard to the prophylactic power of the germs of swine plague.*

Second—*That under artificial conditions of cultivation in any known mode the germs of swine plague do not retain this immunity producing power for any length of time; in fact, that the first culture from the sick hog, and then only a hog selected under very special conditions, would give immunity, or a resistance to natural infection, in inoculated hogs.*

It is rather a remarkable fact that I never failed in producing immunity in inoculated hogs to the most severe tests in exposures to natural outbreaks on farms during my first engagement in Nebraska up to the very last inoculations I made just before leaving the state, and it was from these failures that I first learned that while I had so invariably succeeded, that I was still on the wrong track, though it was not until I had completed and systematized the records of the inoculations of 1891 and compared them with all the records from 1886 to the present time that I thoroughly learned what a narrow margin I had to work on in order to obtain success. This result has been obtained by the inoculation of thousands of hogs and pigs divided into bunches of from 5 to 500 and scattered very largely over the western tates from Ohio to Georgia; Colorado being the most western point. In this review it is only necessary for me to sum up the general result, as it would have no practical value to give the details of each inoculation. Let me again say, *that with the exception of the actual experiment work done at the State Farm, no honest investigation by farmers only desiring the truth can discover any hogs or pigs either killed or permanantly injured by the inoculation of virus sent to the owners directly by me.* I shall speak of this again when I come to consider the assertion of my opponents that inoculation is a "dangerous" procedure on that account.

To return to the

RELATION OF VIRULENCE TO THE PROTECTIVE POWER OF A VIRUS.

When I began my investigations I thoroughly believed that such a relation existed. It has been said that had I diligently studied the reports of the Agricultural Department I should not have made the errors that I have. I have studied those reports probably more diligently and accurately than their authors ever did, and have only learned *that they are the most inconsistent, contradictory, and valueless documents ever published in the name of scientific investigation, and that in no way are they more so than in connection with this very question of preventive inoculation in swine plague.* I have shown up this fact repeatedly, but it is of such vital importance to honest government in this country, and to the swine breeders, that I must be pardoned if I again sum it up in direct connection with the point at issue. The government not only promised a preventive virus in 1883, but advocated its mitigation on the plan of Pasteur as follows:

Our investigations have shown that the swine plague is a non-recurrent fever and that the germs might be cultivated; *they have even proved that these germs may be made to lose their virulent qualities and produce a mild affection.* Surely we have here sufficient evidence to show that a reliable vaccine might be easily prepared, if we carried our investigations but a little way farther. If we had such a vaccine; if it were furnished in sufficient quantities and of a reliable strength; if it proved safe in the hands of the farmers, would not our problem be solved? Could we reasonably expect anything better of this disease?

M. Pasteur has recently confirmed our American investigations in a very complete manner. He has shown that the disease is produced by a micrococcus; that it is non-recurrent; that the virus may be attenuated and protect from subsequent attacks, and he promises a vaccine by spring. (Report 1883, p. 57.)

We now know that every word of those statements and promises were made up out of whole cloth, as neither the government investigators nor Pasteur ever knew they had the germ of swine plague in their possession previous to 1886, and the germ was not even known to exist in France until 1887. Pasteur worked on rouget and not swine plague, and did not discover the germ of that disease, which honor belongs to Loeffler and Schütz in Germany, and was made in 1885.

THE GOVERNMENT ADVISES THAT MITIGATION OF VIRULENCE IS UNNECESSARY IN A VIRUS TO PREVENT SWINE PLAGUE.

It says:

We soon found that there was no indication of attenuating the virus for this purpose, because the strongest virus might be introduced hypodermically with impunity in considerable doses. Now as the stronger a virus is the higher degree of immunity it produces, there is every reason for using the fresh unattenuated cultures. *But even these are not sufficient.* (Chief of Bureau of Animal Industry in *Journal of Comparative Medicine*, 1888, p. 148.)

I have most carefully examined all the reports of the Agricultural Department from 1885 to date, *and cannot find one case where "a fresh unattenuated culture" has been once used and its use free from most decided objections in their attempts to prevent swine plague by inoculation.* (By "a fresh unattenuated culture" I mean one made directly from a recently attacked hog and from a fresh outbreak, and used within seven days from the time the bouillon was inoculated, and not one used in a generation made from such, or from a culture passed even once through any small animal. This is one of the vital points in preparing a virus which will prevent, if used right, and the hog from which it is taken correctly selected.)

MITIGATION PRACTICED IN THE FACE OF THE ABOVE QUOTED POSITIVE STATEMENT THAT IT WAS TOTALLY UNNECESSARY.

The first steps in this direction (of preventive inoculation) were made at the Experiment Station early in 1886, soon after the hog cholera bacillus had been discovered. (Report 1890, p. 110.)

How about those experiments which "*even proved that these germs may be made to lose their virulent qualities and produce a mild affection,*" reported in 1883?

Which is that, consistent or inconsistent? The public would like to have the secretary of agriculture answer that question without prejudice or bias.

We soon found that there was no indication for attenuating the virus, because the strongest virus might be introduced hypodermically with impunity in considerable doses. (Report 1888.)

Another striking example of the consistency of Mr. Rusk's authorities is to be seen by comparing the above with the following:

The vaccine used: In order to obviate the fatal effects of doses of hog cholera cultures injected under the skin, which sometimes shows itself quite unexpectedly [They cannot show one such a case in all the experiments which any scientists would accept as free from objections as noticed by Frosch], especially in young animals, the writer deemed it advisable to reduce the virulence of the cultures by appropriate means, so that larger quantities of the culture liquid might be injected to increase, if possible, the vaccinating effect without endangering the life or stunting the future development of the animal. [Which advice shall we follow, that of 1888, which positively tells us that attenuation is not necessary, or of 1890, which says it is?]

In reducing the virulence or attenuating it, the following method has been pursued: Tubes of bouillon inoculated with hog cholera bacilli were placed in a favorable temperature for multiplication (95 to 100°) over night; on the following day the culture liquid, now slightly clouded, was placed in an unfavorable temperature of 110 to 111° F. and kept there for ten days. Thereupon fresh tubes were inoculated from these and subjected to the same process. From time to time rabbits were inoculated to test any attenuation that had taken place, and it was noticed that there was a slight modification of the disease in rabbits after a time. After the bacteria had been thus exposed to a high unfavorable temperature *for more than two hundred days, and passed through twenty cultures, a small dose of one-tenth cubic centimeter injected under the skin did not prove fatal to a rabbit while larger doses still were.* The reduction of virulence was not, therefore, very great even after this very prolonged exposure to high temperature. At the same time it was thought advisable to use it as "vaccine *a*."

A second vaccine was prepared at the same time. It was exposed for only ninety to one hundred days, and passed through nine cultures in place of twenty as with vaccine *a*. This we shall call vaccine *b*. (Report 1890, p. 111.)

Will or can any competent investigator, or any one with a grain of common sense, tell why such a procedure should have been gone through with if what the chief of the Bureau of Animal Industry said in 1888 was true or based on the results of one single experiment, viz., that "we soon found that there was no indication for attenuating the virus, because the strongest virus might be introduced hypodermically with impunity; and, as the stronger the virus is the higher degree of immunity it produces, you can see that there is every reason for using the fresh unattenuated cultures"? If that is not being "inconsistent" the dictionaries and general understanding of the meaning of that word are certainly most seriously at fault.

That all attempts at producing protection with such an "attenuated virus" should fail must be self-evident if the words spoken in 1888 were true. Any one who will carefully read the reports of the department on swine plague or who has experimented with this germ in order to prove its specific etiology in swine will soon learn two things:

First—That the words spoken in 1888 were true as to any general danger of fatal results. Only cultures from the most exceptionally virulent and very fresh outbreaks are dangerous to young, healthy pigs over three months old, if allowed to run free, in doses of one cubic centimeter; where fatal results are desired one of the best ways to assure them is to pen the animals up closely so as to give them as little movement as possible.

Second—That the germ loses very fast in virulence in pigs if sick long, and that it can be robbed of all virulence by being passed through a series of pigs, and dies out in diseased hogs in a few weeks.

The government investigators have been perfectly aware of these facts in the past and have borne witness to them, but they completely forget both when it becomes necessary to confuse and mislead the public with regard to our Nebraska work. In the report of 1885 they say:

> *The fact that it is difficult to demonstrate the presence of the bacterium of swine plague in chronic cases which have lasted more than three weeks and in which the ulcerations in the large intestines are already far advanced, cannot be emphasized too much.*
>
> Ignorance of this fact has no doubt led to previous erroneous deductions in investigations on the etiology of this disease. *Chronic swine plague must henceforth be looked upon as an after stage, independent of the disease itself,* and caused by intestinal lesions, the indirect result of the growth of the bacterium in the blood vessels of the mucous and submucous tissues. *The bacterium has already disappeared from the stage,* and makes way for either harmless or septic microbes which gain entrance through the ulcerated intestines and are found in the blood and serous exudates. (P. 211.)

The stage at which "the bacterium has already disappeared" has been given as "chronic cases which have lasted more than three weeks."

Every word of the above quotations is generally true and will be confirmed by every observer who may in the future investigate this disease. Such cases must be carefully avoided in the selection of a hog from which to obtain virus, yet they are just the ones the majority of farmers are inclined to send to a laboratory, simply because they are emaciated and useless hogs. When the germ is present in such cases it is often excessively mitigated in virulence for pigs, as well as entirely without any prophylactic properties. This does not apply to rabbits, which are especially susceptible to the injection of this germ; infinitely more so than small pigs even. It is wonderful what psychological somersaults the investigators of the government have forced themselves to make in their malignant opposition to the Nebraska investigations, and how recklessly they throw all consistency to the winds between one report and another. Keep the above in mind and read this from the report of 1891:

In case of hog cholera we have found the bacilli in the organs of swine *six to seven months* after apparently unsuccessful inoculations. *These bacteria possessed the original virulence.* (P. 123.)

How could that be if the inoculated swine were apparently well and the inoculations had been "apparently unsuccessful"?

Surely the "original virulence" must have been very near null in such a case as that. The "original virulence" was certainly tested by rabbit inoculations, which are to be banned and condemned as worse than useless, because misleading.

A HOG OR PIG SICK OVER A WEEK UNSUITABLE TO MAKE VIRUS FROM.

Why this is so we do not know. We simply know the fact from many experiences. It is probably due to the germs having, through changes in nutrition offered by the porcine organism, lost the power of producing a sufficiency of that chemical product which enters into the unknown chemical combination with the product of some of the cells of the infected organism, which gives to it the power of resistance to a second infection in non-recurrent diseases. What these products are, how they combine, and, perhaps, how resistance to second infection is produced, are all questions for the chemist *per se*, rather than the pathologist, for the work of investigation of infectious diseases is opening so many new and special lines of investigation as to call most emphatically for more and special investigators in these subdivisions of labor. The pathologist can no longer encompass the whole field of pathological research. It will be remembered that in the report of 1885 the Agricultural Department told us "*that it is difficult to demonstrate the presence of the bacterium of swine plague in chronic cases which have lasted more than three weeks.*" Much fault has been found with the method which I use in obtaining the virus for inoculation as we use it in Nebraska, because in the practical work, not that of investigation, I affirm that the exact methods of the laboratory are unnecessary; by which I mean plate and control cultures. Not only do I assert that they are unnecessary, but that the practical result has conclusively demonstrated that the less we interfere with a culture, the less we carry it under artificial conditions, the quicker we use it after making it from a diseased hog, the more certain are our preventive results in the animal inoculated if exposed to infection. The shorter the time the germ has been in the animal, the quicker we obtain it after infection, the more reliable is the culture to give the inoculated hog resistant power towards natural infection. Again, as the government also well said in 1885: in these chronic cases, which have "lasted over three weeks," "the bacterium has already disappeared *and made way either for harmless or septic microbes,*" which gain entrance through destructive disturbances set up previously by the specific germ. To avoid all these disadvantageous points, I soon found out:

First—That the most satisfactory and easiest way to get pure cultures was to kill a hog or pig as soon after being taken sick as possible, and to use the heart's blood or spleen for that purpose. After killing the animal it is always well to allow it to remain in a cool, shaded place for thirty minutes or more to allow local development of the germs after the circulation has stopped. This avoids, as far as possible, the formation of destructive lesion, which act as *ostia* to adventitious germs. In all my work, hogs being plenty, I have been enabled to follow this course, and to use dead hogs, or those sick longer, and hence more or less emaciated, for the study of the lesions in the disease, both together, permitting a close following of the lesions from the primary or specific to the secondary complications. During my earlier investigations, the most exact attention to plate cultures was given, and thus I learned the conditions which were accompanied by the presence of adventitious germs, and by selecting freshly attacked animals avoided their presence. As I have said in another place, small pigs being cheaper and easier to be had, and as I was studying swine diseases, I had but little recourse to rabbits in the first year of my investigations.

For practically obtaining, in nearly every instance, a pure culture of the germs of swine plague the above course will do, but to obtain a reliable virus it is not sufficient even to take a hog just taken ill; it must not only be a hog just attacked, but the outbreak must have only just begun; in fact, where possible, *those who desire to succeed must make it a rule to select outbreaks that have not run over one week and a hog or pig just taken sick.* The reason of this is, that that insures as positively as we can do it that the germs have not been in the porcine organism but a few days.

SWINE PLAGUE NOT A CONTAGIOUS DISEASE.

It is very well known that the government investigators, among their almost innumerable contradictions of nearly every conclusion which I have published, also contradict me when I say swine plague invariably finds its primary origin in conditions outside the pig organism; in other words, that the germs are a natural inhabitant of certain soils: they say that *swine plague is a contagious disease*, which I as positively deny. Those who are acquainted with the opinions I have determinedly defended know that by a "contagious disease" I mean *one in which, as far as we know, the germ finds its primary locus of development in some species of animal life, and never, primarily, in outside conditions*, like glanders, syphilis, contagious pleuro-pneumonia, etc. I affirm that any form of transmission between diseased and healthy individuals, no matter by what means, direct or indirect, natural or intentional, accidental or experimental, has nothing whatever to do with deciding that one essential point. That this class of diseases are endogenous, as Pettenkofer calls them; or "obligatory parasitic," to speak with Hueppe; *that is, in order to continue to live, the germs causing such diseases are absolutely dependent on the tissues of some species of animal life for nutrition under natural conditions.*

Now I assert, with dogmatic positiveness, that any person who dares assert that swine plague is a disease of that kind is an ignoramus, and a disgrace to that intelligence which should be the distinguishing attribute of a member of the medical profession. I am perfectly well aware that I am hitting some tolerably imposing (in their own minds) authorities when I assert that; that many of the most noted men in the British medical and veterinary professions, the teaching of the British schools, the laws for the suppression of swine plague in England, and the veterinary authorities of the Privy Council, all assert swine plague to be a "contagious disease," and yet in the face of so much august authority, *I tell them one and all that they do not know what they are talking about when they say so.* They may kill off all the swine in Great Britain and pay for them out of the public funds; they may forbid the raising of swine in all Britain for three, five, or ten years and buy all their pork from foreign countries, and then I tell them, with equal dogmatism, *if they stock their farms again they will have swine plague again within one year, and they may buy assuredly healthy stock, and every hog bought may be well for three months after landing on British soil, which will certainly forbid the animal having been infected when imported.*

The imbecilic, idiotic idea that swine plague is a contagious disease and can be stamped out by killing as is done in "contagious diseases," such as pleuro-pneumonia, glanders, etc., is also advocated and "published by and with the authority of the secretary of agriculture" of the United States, the great and only authority on such questions, to whom the agriculturists of this great and free country bow down to as some wonderful and pan-omnipotent fetich. The following is one of the latest promulgations from the throne of agricultural (dis)grace:

HOG CHOLERA.—SECRETARY RUSK'S VIEWS UPON INOCULATION.

In answer to an inquiry from the *National Stockman*, Secretary Rusk writes as follows:

I take no stock in it whatever, and I go so far as to say that if the investigations of science should demonstrate the efficacy of inoculation for this disease, it would in my opinion be a practical failure, even though it might be termed scientifically successful. In the first place, to keep the farmers supplied with virus means to perpetuate the disease somewhere or other all the time, and that I don't believe in. Again, inoculation, to have any effect at all, means two weeks' sickness of greater or less severity to the victim, then two weeks more to get over the effects—I refer, of course, to fattening hogs—and you have thirty days without gain. In my opinion, considering that we now market our hogs at about ten months from birth, the loss of this one month means the loss of profit on the animal. This is, mind you, without taking into account the fact that inoculation, if performed at all, must be performed by a skilled professional; no other person should be allowed to handle such dangerous matter as hog cholera virus. As to this idea which some people advocate, that if inoculation is found to be efficacious, any farmer or farmer's hand can be instructed to extract virus from diseased hogs and inoculate the healthy ones with it,—it is the wildest notion I ever heard put forth, and I cannot see how any sane person can entertain it for a moment. [Mr. Rusk should read his own authorities and advisers' letters on this subject, in one of which it is said that any one can make the virus. Mr. Cadwell made a culture of the germs at Ottawa. Read: "And in the reports of this bureau for 1886, pages 60 to 70, the details of the experiments are given so fully that any one who can make a culture of germs can repeat the experiment himself."—Letter of Chief of Bureau to Editor of *The Homestead*.] The idea of handling the virus of a most contagious disease—the spread of which means many millions of dollars loss annually to the farmers of this country—without the utmost precaution and under the strictest supervision, is preposterous. For my part, as I said at the beginning, I would oppose the inoculation plan, if only because it necessitated a perpetuation of the disease, under any circumstances. What we want to do is to treat it as we did pleuro-pneumonia among cattle—wipe it out.

If asked, Why is it that the subject has continued to receive the attention it has from the officers of the Bureau of Animal Industry? That is easily explained; while I have no confidence myself in the practicability of inoculation, I soon found after assuming my present office that there were conflicting views in the country regarding the subject. I determined that the most effectual way of putting an end to the controversy was to permit Dr. Salmon to continue the investigation undertaken in the bureau to the end. You can't satisfy people by telling them simply you know a thing won't work; you must be able to show them that you have tried it and tried it thoroughly, and prove to them that it won't work. Consequently, while the chief of the bureau himself declared that inoculation had not been brought to such perfection as to justify the anticipation of favorable results, I instructed him to satisfy the farmers of La Salle county by participating, through a representative specially assigned for the purpose, in the inoculation experiment they desired to carry out, and you know what the results were. We will shortly publish a bulletin, brief and in plain terms, on the subject of inoculation, and I expect it will be sufficient to convince everybody that, whatever scientific men may be able to accomplish in the way of preventing hog cholera by inoculation, the system can never be practically applied without destroying the profits on a man's herd, and we will offer ample evidence from the practical experience of those who have tried it that the man who takes the risk of experiment on his own herd with hog cholera virus, is, to use a slang phrase, "monkeying with the buzz-saw."

Such advice is criminal in more ways than one. Mr. Rusk is so totally ignorant on all such questions, and thereby absolutely unfitted for the responsibilities of the position which he incumbers, that he is led in any way it pleases the chief of the Bureau of Animal Industry, who, as he is Mr. Rusk's instructor, is evidently as ignorant as his most pliable pupil.

Let us see why it is absolutely certain that swine plague cannot possibly be a contagious disease. All contagious diseases which we know of have always occurred in new countries or districts where they have not existed most generally through the admittance or entrance of a diseased individual from some infected locality, and much less frequently by means of material of one kind or another directly emanating from such an individual. In other words, we have no record of the origin of obligatory parasitic (or contagious) diseases in any other way. For us a diseased individual must have been present. Now, according to what we know, swine plague first made itself so apparent as to attract attention somewhere between 1830 and 1840 in this country. No infectious disease degenerates *de novo*—out of nothing. The cause must have been either here, or been introduced here from some country in which the disease existed by sick swine or manure, or something directly from sick swine. Even the government contagionists have never hinted at the importation of this disease from Europe. It would be ridiculous to do so in the light of conditions existing fifty years ago. Sick swine would not live long enough to have been imported alive in the slow ships of that period, and the landing of offal or manure from such swine is too utterly improbable to be thought of. It must not be understood that I in any way deny the possibility and frequent occurrence of the conveyance of swine plague from one place to another. I only say that it would have been impossible under conditions existing at that time to have introduced it from Europe. We know that uncleansed stock cars have been the chief means of extending the disease in the states west of Ohio, where some say it first originated in this country. As the disease could not have been imported and could not have originated *de novo*, out of nothing, the cause of it must have existed and been in the country somewhere.

A prime factor in the cause of swine plague which does not seem to have attracted the attention it deserves is the nature of the soil of the localities in which it is in reality a plague. It is not a disease of New England, or of our northern states, unless imported, and then dies out of its own accord or without much attention. The line of northerly extension in Wisconsin is quite sharply drawn by the change of the soil from a rich, heavy loam to a light soil with gravel or sandy subsoil. I understand, or am informed, that the same is true of Britain where the disease exists and where it is seldom, or does not permanently establish itself. There is not an obligatory parasitic disease on earth (a contagious one in the only sense the word should be used, not that of transmission but primary origin) that is thus limited by climatic or telluric conditions; though the former may and frequently does exert favorable or unfavorable conditions towards their extension when introduced; the confinement of cold weather favoring and the freedom of the open air in summer being less favorable to their rapid extension; in other words, the close grouping of individuals. I would like to see the Washington authorities come out squarely and assert that swine plague is not a faculatative, parasitic, infectious disease. They dare not, for they know that every scientific investigator in the world would be against them. In calling this plague "contagious" they are taking refuge behind the ignorance and carelessness of the medical profession in the logical use of medical terms, which uses the word contagious according to its root-meanings as coequal with transmissibility; but that is not and never has been the practical meaning attached to the word, for we would not find the absolutely universal expression of common, every-day experience that "*while all contagious diseases are infectious, all infectious diseases are by no means contagious.*" As both classes are transmissible in one way or another it must be evident that there must be some one essential reason for such a well-recognized differentiation. That one reason is the source of primary origin of the cause. In the one case, the contagious, or obligatory parasitic diseases, it must be and is in the diseased individual; the germs only finding in such their proper home and food—*they are obliged to be parasites.* In the other, the germs, the cause, live naturally and primarily in conditions outside the animal body—but can live in certain animal bodies and there do produce infection, because the food offered is favorable to that effect. Hence, they are exogenous diseases; that is, they have the faculty of living and being infectious in certain forms of animal life. They are faculatative parasitic diseases. Swine plague is such a disease. Its germ must be a natural inhabitant of certain soils under certain conditions, but only where those soils have become saturated with the excreted materials of swine do they change their character and become toxic, but only naturally to swine, because the tissues of swine offer them the same toxic-producing food; hence, swine-plague only in certain localities is a permanent or oft-appearing infection. It is nothing contradictory that the germs lose their virulence in time in swine, for unless they did none would recover, and even the obligatory parasites for a time do the same thing, but with this difference, that swine plague germs do not regain it in passing through swine, while obligatory parasites do in the species in which they naturally cause disease, if continuously passed through a succession of fresh susceptible individuals. Were the swine plague germ naturally infectious and universally distributed, and were not its toxic effects produced through saturation of the soil with swine materials, we should know it as a frequent infection in rabbits, tame ones, as they are manifestly more susceptible than swine, but this is not so. We shall return to this subject shortly.

Again, swine plague, uncomplicated, when studied in its earliest stages, is not necessarily either a pneumonia or an ulcerative or neoplastic inflammation of the intestines, *but a septicæmia. There is not a septicæmia known to man the cause of which is not faculatative parasitic.* Not one in which the germs thereof do not originate primarily in extra-organismal—outside—conditions. It is now years since the Agricultural Department first asserted this disease to be contagious. Why not more work and less talk? Why not have long ago taken some severely infected western farms and proved their assertions by actual demonstration, by buying the hogs and paying the men not to keep hogs for some years and cleaning off the refuse and manure once thoroughly; in fact, giving such places exactly the treatment done in contagious pleuro-pneumonia, and then leaving the yards for three years unoccupied by swine and put new healthy ones on. If there is money enough to squander in a bitter personal fight against the work in Nebraska there should be funds enough for such a demonstration. It would do no good. I can quote case after case where almost that very thing has been done; where farms have been vacated for several years and given up to banks because the owner had been rendered bankrupt by swine plague; where no human being or live stock had been on such farm for several years; where the banks have either let or sold them, and on being restocked swine plague has soon appeared in a most virulent form, and in hog yards exposed to the vicissitudes of the weather for several years with no buildings to shelter or protect the germs on them. No obligatory parasitic or contagious disease is known to man that works that way. I know of cases where the transportation of the disease to such farms is absolutely excluded, there being no other farms near enough and no other disease in the vicinity; such restocked farms being the first and only case to occur for a long time in that section of the country. Every observant hog-raiser can call to mind similar cases. More are known where men have given up all hog raising for a series of years and then restocked, to have the disease occur in their newly purchased and assuredly healthy hogs when there were no other cases in the vicinity. Contagious diseases do not act that way. It requires only a little close thinking on the part of swine breeders to demonstrate to themselves the unquestionable correctness of my views and the incorrectness of those advocated by the secretary of agriculture.

It is more than probable that Mr. Rusk has been instructed that swine plague is a contagious disease by his teacher, because, were that view to be practically applied in the way it is necessary to do in such diseases as contagious pleuro-pneumonia, it would necessitate the employment of hundreds of political heelers and wire-pullers who would be most useful tools in elections to those desiring a mortgage on public positions.

If Congress means to be honest it will be a long time before it saddles any such useless and expensive fraud on the backs of our altogether overtaxed farmers. It must be either preventive inoculation or cure, or both, and it is the duty of our government to intelligently and honestly seek and perfect them by continued investigation. The one has been conclusively demonstrated to be a fact; the other promises to be possible and is more than probable.

LET US RETURN TO THE CAUSES OF FAILURE IN INOCULATION!

We left the subject after noticing the kind of hog to select in order to obtain virus. Let us consider now the kinds most likely to be sent in by farmers or most willingly given for that purpose. No one unacquainted with all the facts can realize what a vast amount of actual experimentation has been done in order to arrive at the knowledge we now possess. It has been one continued series of experiments since 1886, experiments greater in number and covering more individuals than almost anything else of the kind, unless it be with Koch's tuberculin.

The hogs on the farms have been used in the experiment, besides hundreds that I have had a personal interest in. A record has been kept of every hog used for inoculation, with as complete a history of the outbreak as possible, and a record of the pathological lesions in the animals used to obtain virus from. Take the year 1891, for instance; forty-six bunches of hogs were inoculated, covering nearly 3,000 animals, the virus being taken from such hogs as the farmers sent in, in order to see how closely they would follow directions. These forty-six bunches varied in numbers from 5 to 297 animals. What experiment station could supply the means and place to make such a test as that? It would be impossible; and still more so to have inoculated all these bunches at the state farm, kept them separate, and then sent each bunch out into outbreaks to test them. I do not desire to sound my own praises, but must say that few men in my position, with the continued distrust fathered and continued by the Agricultural Department, would have had the self-confidence and courage to have gone boldly out on the farms and done as I have. I have boldly placed my reputation in the balance and used virus that I had evidence would fail, and inoculated hogs on farms which I was almost sure would not be protected, in order to conclusively demonstrate a valuable point which I had not means enough to do at the experiment station. That kind of work is about done, as this publication will show. About all the reward I have thus far received has been wholesale condemnation from those who should, in justice, have kept their "hands off."

As has been said, no honest investigation can find reliable evidence of a single animal having been injured by inoculation itself. It has also been asserted that we had no failures up to the time we left Nebraska in June, 1889, though we did not then know why we succeeded simply because we had not failed. Mind, *I positively deny both failure and injury in hogs inoculated at Surprise, Neb., in November, 1888*, because both those that were not sick at the time at Surprise, Gibbon, and the state farm were as severely tested as hogs should be, and always in company with healthy uninoculated stock subsequently. No one of any sense would think of denying that inoculated hogs cannot be killed by overcharging them with virulent germs (or that inoculation may sometimes fail in extra virulent farm exposure, or from insufficient inoculation, or unsuitable virus) as was done by authority of the Department of Agriculture, in absolutely unnatural experiments with some of these Nebraska hogs of 1888, but the person who made such nonsensical and unequivocally unscientific tests himself says:

The last two series of experiments show that whatever the force of the immunity may have been, there was an artificial means of overwhelming its protective power. *Such experiments do not, however, conclusively prove that there is no immunity, either naturally or artificially acquired, capable of practically protecting against a natural attack.*—(Report Bureau of Animal Industry, 1889–90, p. 143. E. O. Shakespeare.)

The latter part of that statement is supported by words in Dr. Salmon's letter to the editor of the *Homestead* as follows:

It is possible, however, as I have freely admitted, though hardly probable, that the conditions of exposure on farms are not so severe. In that case a degree of protection, which in our tests was insufficient, might ward off the disease in most cases on farms.

What on earth, then, is the use of wasting public money in experiments that have no practical value? To the ordinary farmer, however, they would certainly serve the purpose of creating distrust in our Nebraska work, which was the sole reason that they were attempted. All the farmer, if he read such stuff, would see is, that the Nebraska inoculated hogs were killed by these unnatural injections of large quantities of the most virulent germs obtainable directly into the blood in one dose, and not taken in at intervals as in natural exposure in an outbreak; their entrance to the blood being also obstructed and retarded by their having to pass through various membranes. The farmer would have probably forgotten, if indeed he ever knew, that this same Shakespeare did, in what pertains to be his report as commissioner, say that "*the Nebraska (inoculated) hogs stood the test better than recovered (from natural disease) hogs,*" and the coadjutor of this man, in his attempts to deceive the public and his unscientific investigations, and shirking the most solemn obligation a man can have, Prof. T. J. Burrill of the State University of Illinois, also said, even more emphatically, the same thing in the following letter to me:

CHAMPAIGN, ILL., Aug. 10, 1889.

DEAR DOCTOR: Have just answered your telegram to the best of my ability. When I saw the pigs no trial had been made upon the Nebraskans with Salmon's "swine plague," and I am quite sure nothing was said about it to me by Dr. Shakespeare subsequently. *Abundant trial had been made, both by feeding and by subcutaneous inoculations with the Washington "hog cholera, without serious effect. None stood the test so well as those Nebraska pigs.* It is, however, quite possibe that Dr. Shakespeare has tried "swine plague" (Salmon's) since our correspondence upon the subject. Have not seen anything in print on report. Am not certain what publicity has been given. The two diseases are acknowledged, the one, however, said to be much more prevalent than the other. No attempt is made in the report to compare European work, though we did have some cultures from abroad.

You can understand that it was difficult for separate workers to make a report. Would liked to have published full details but found only certain conclusions could be agreed upon, not always worded as one would do it for himself. At any rate, nothing was considered except what is the fact.

Very truly yours, T. J. BURRILL.

The chief cause of failure has been, and will be, that the sick hog, from which the germs for virus will be taken, has been sick too long.

Remember, the hog for such purpose should be selected from an outbreak that has not existed for over one week, and if possible, be the last pig to show signs of illness in such an outbreak. Inoculation should be done twice to make surety sure. But in my earnest desire to make it as cheap as possible to the farmer, I have tried to do it with one inoculation, when the cost would not be over fifty cents for 1,000 hogs. But, while successful in the majority of cases, when the virus was right, it is not safe to rely on one inoculation. We cannot get in enough germs to thoroughly saturate the porcine organism, as is done in the fractional infection of those mild natured outbreaks which give the most absolute immunity, but with a second inoculation of rightly selected virus we can give doses from one to three cubic centimeters the second time with perfect safety, according to the age of the animals.

The government has told us that it was only with great difficulty that the germs could be found in cases of over three weeks old. That is not quite correct. The germs can generally be found, even in such cases, if care is taken to isolate them, but not always in a pure condition, being often mixed with numbers of adventitious germs.

We had failures, after moving to Chicago, using the virus made from the germs of such hogs, but from none that had been ill so long as three weeks. When we came back to Nebraska we concluded, as farmers would be most likely to to send us just such hogs, to make one direct and extensive experiment in that direction. The government says that it has reliable information that we killed certain pigs by inoculation at the state farm in 1891. All I have to say is this: That all the work I do at the state farm is experimental, and that such experiments are the property of this laboratory until we get ready to publish them, unless the results are demanded by those having the authority to do so, though all records are free and open to the public. The "three different correspondents of undoubted reliability," whose report appears in Bulletin No. 8, are three cowardly sneaks, whoever they may be, and their report as unreliable in spirit and as false in fact as the majority of the testimony in that bulletin. Why not publish the names of such undoubtedly reliable correspondents? The government was afraid to. It is more than probable that their reputation for "undoubted reliability" is more than doubtful in this community. Here is the whole passage from Bulletin No. 8:

FAILURE OF INOCULATION IN NEBRASKA DURING 1891.

Under date of January 6, 1892, Dr. Billings addressed a letter to the Omaha *Bee*, in which he endeavored to explain the communication of disease by the inoculations made in accordance with his method at Ottawa, Ill. The following extract from his letter is of interest in this connection:

"I have inoculated some 50,000 hogs, and never in a single instance that I know of has such an accident occurred through inoculated hogs as at Ottawa, and there have been very few cases in which inoculation has not protected. True, I failed completely in protecting hogs that were fed on glucose refuse, except those from Surprise, Neb., but that was due to the glucose and not the inoculation. Hogs fed on distillery slops can be protected by inoculation. Every one who is acquainted with the true facts knows that those herds reported as killed at Surprise, Neb., in 1888 were all diseased at the time they were inoculated. This year over 3,000 have been inoculated in Nebraska, and to-day I sent out virus for 1,900 more, but with some regrets, as I fear for its injury and the possibility of its being frozen. Of the 3,000 I do not know of one being injured by inoculation, yet one such case in sucking pigs is reported, and one failure in the same herd; the pigs I doubt, as five other lots of pigs were inoculated at the same time with the same virus, and they all lived; the failure I know the cause of, and have learned to avoid it in the future."

In spite of this very positive statement, the department is in receipt of information from three different correspondents of undoubted reliability to the effect that on the 12th of August, 1891, forty-eight head of swine were inoculated on the state farm under the direction of Dr. Billings, and four of the herd were not inoculated. August 30, four pigs were dead, and two others very sick were taken to the laboratory for examination. Within thirty days after inoculation, twenty-six died, and before the outbreak set up by the inoculation ceased its ravages, forty-one of the fifty-two hogs on the farm died. These facts were certainly known to Dr. Billings at the time the letter quoted from above was written.

With *Western Resources* for February 10, 1892, was included a supplement giving a statement by Dr. Billings of the inoculations made in Nebraska from August 18, 1891, to January 1, 1892. Why the inoculation on the state farm of August 12 was not included was not stated.

In his open letter to Chancellor Canfield, Secretary Rusk hauls the superintendent of the state farm over the coals for not reporting this experiment. I will say to Mr. Rusk that it is not the method of gentlemen to go to servants for information. Neither Mr. Perrin, the professor of agriculture, who is really the head of the state farm, nor even Chancellor Canfield has anything to do with the experiments carried on by this laboratory, no more than I have with their work. What would he think of any one outside who desired to know of the work of his experiment station who should apply to the foreman of his men? Experiments are the property of the experimenters until published, or until those having the right demand the results. Such a demand had not been made on me until Bulletin No. 8 appeared, when I reported to the chancellor. The affair at the state farm was an experiment, and not an inoculation to prevent, but to see would it not fail? Mr. Perrin reported honestly on the results of inoculations to prevent. A great many pure-bred stock hogs have been sold from the farm on the warrant that they were inoculated and would withstand farm-exposure. All we know is that no purchaser has ever applied for the return of his money; which is sure evidence of satisfaction, as men are not over scrupulous in making demands on the public funds on slight excuses.

Had those "three very reliable correspondents" been honest citizens of Nebraska they would have come to the laboratory and made some inquiries and found out the particulars about that inoculation at the state farm, and would have discovered that not a hog or pig died from the effects of the inoculation; *that the virus was not expected to protect, and that just what we hoped for happened, though we expected to have to turn the hogs out into some outbreak to attain it, but were spared that trouble and expense by the disease* breaking out at the state farm.

WHAT THE PROFESSOR OF AGRICULTURE REPORTED ABOUT IT.

The following extract is taken from the report of the professor of agriculture to the regents of the university: "The hogs upon the farm have been used for purposes of experiment in inoculation, and quite a large number succumbed to the disease. The loss to the department was considerable, but the result to the state, no doubt, is worth the price of many herds of hogs in the establishing of a scientific truth for future guidance of the farmers of Nebraska."

C. L. INGERSOLL.

All Professor Ingersoll had to report was what became of the hogs raised. The balance of the question belongs to this laboratory.

In the same letter to Mr. Canfield, published at the end of this work, Mr. Rusk says:

I might have applied for information in regard to inoculations made in Nebraska as you intimate, but if the malignant hostility shown toward this department by the person in charge of your laboratory had not prevented me applying to that source, the unreliable statements made from time to time, issued by him, would have been sufficient. Comparing his statements made from time to time with letters received from unbiased citizens of Nebraska, I have no hesitation in saying that even if the latter were made from memory they deserve the greater confidence.

The value of the "confidence" to be placed in the "memory" of Mr. Rusk's "unbiased citizens of Nebraska" has been fully demonstrated. It is begging the question to say that my "hostility" prevented his asking for information. He never asked. He would have got it, with all the details from the letters of farmers in full had he asked, entirely regardless of my personal feelings in the matter. Not a false or even misleading statement has ever been sent out from Nebraska intentionally, and I do not know of one being sent at all. The evidence here given in detail proves that. The long expressed confidence of the best breeders of the state proves it. There is no clique of a "few interested persons" holding me up. I am desirous to leave, but cannot do it in justice to the people and shall not until I am turned out, or I can no longer do them justice. There are no financial charms to keep me in Nebraska. I can earn or have a larger income in other ways which my duties here prevent me from obtaining. I am here to gratify my ambition to serve the people. So long as their desires coincide with my own purposes in life I shall stay, and no longer. That is, so long as in serving them I am serving humanity and the cause of original research.

It is very singular that such "reliable correspondents" should have neglected to mention a much more important fact in that connection, which was, *that not one of the hogs inoculated previous years were sick, though on the same ground and among the sick hogs.* This is illustrative of the whole manner in which Mr. Rusk and his agents deported themselves in obtaining information regarding the work of this laboratory. It has been shown that their records in Bulletin No. 8 are not only false in many respects but generally unreliable, and that a very large amount of truthful evidence favorable to our work has been suppressed. This is a public laboratory, not supported by the people of the state of Nebraska, but by the citizens of the United States, of whom they are but a small part. Everything we do in our official capacity, every record here kept, is constantly open to the public and can be seen by any one. There has been and is nothing hidden. It was my intention, and it may be done yet, to publish exact statistical details of all my work on inoculation, but, as has been said, it is now only that the proper time has arrived. Mr. Rusk could have selected any of his correspondents here, such as Judge Eaton or ex-Governor Furnas, and sent them here, and they would have had every opportunity to get at all the facts without suppression of any; but instead of that he preferred the underhand and dishonorable method of obtaining his information generally from the people who knew nothing of the subject. From those who knew something he has published little or nothing unless they said something which could be construed as unfavorable to our work.

Now, as regards those pigs inoculated at the state farm August, 1891. I desired to demonstrate that if a sick hog had been ill two or three weeks that it would be useless in producing prevention. There was no swine plague around Lincoln that I could hear of at the time, and I had demands for virus for about 1,000 head and desired to send something in order not to have it said that I was afraid to send it out, as would have surely been the case had I tried to explain that I could not until I got a certain kind of hog. On August 8, 1891, I heard that there was disease at Diller, Neb., and went there, but could find none but one outbreak that had been dallying along for some weeks, and all the pigs were sick and had been for some time;

it was a very slow outbreak. Though unsuitable for virus I determined to kill one and use it experimentally, and if I found the germ sufficiently pure to send it out and instruct the farmers that I would send a second virus in two or three weeks in order to satisfy the demand. The pig selected had chronic pneumonia and chronic intestinal lesions, and had been ill some three weeks and was very much emaciated. Cultures were made and two rabbits inoculated with the blood from the heart mixed with sterilized water. One rabbit died in about fifteen hours, and from its heart's blood was obtained the government swine plague organism. The other did not die. The cultures were hurried up in the thermostat, and some contained a mixture of the government germ and the true swine plague germ, while some contained the latter pure. These were subjected to plate tests and no other germs were present. As I have never had any occasion to see any cause for alarm in subcutaneous injections of the government germ in swine after a great many tests since I first came across it in Chicago (let me say I never had the thing in my cultures during my first work in Nebraska, as I shall show elsewhere, all assertions to the contrary) I determined to inoculate all these pigs and hogs with a mixed culture containing both germs, and the following bunches were inoculated with such a mixed culture:

State farm	48
John Morgan, Lincoln, first inoculation	17
Jacob Dick, Lincoln, first inoculation	5
J. P. Cully, Diller, first inoculation	44
A. M. Caldwell, New Holland, Ill., first inoculation	115
Joe Antes, Syracuse, Neb., first inoculation	50
F. H. Connelly, Dorchester, Neb., first inoculation	163
*E. E. Pamblade, Axtell, Neb., first inoculation	73
Wm. H. Diller, Diller, Neb., first inoculation	98
Spencer Day, North Bend, Neb., first inoculation	70
†Gustav Lundren, Axtell, Neb., first inoculation	77
Dr. H. G. Leisering, Wayne, Neb., first inoculation	50
F. I. Foss, Crete, Neb., first inoculation	153
Matthew & Jenner, Loup City, Neb., first inoculation	31
	994
Less the state farm hogs	48
	946

* Second inoculation, a second generation.
† Second inoculation, a second generation from one of the farm pigs.

On or about August 24 these 946 all received a second inoculation of a properly selected virus except those at Axtell, Neb. (see foot notes), and nearly all, with that exception, stood quite strong tests. The loss at Axtell in surrounding herds was 95 per cent, and in those inoculated less than 50 per cent. *Not one of the 946 was sick at all after either inoculation, nor injured, as shown by letters previously given.*

In less than two weeks after the farm pigs were inoculated they began to come down sick, with the result reported to the government. The virus was useless, as I knew, but desired to demonstrate in this case conclusively, by actual experiment. Did even the mixture kill the pigs, or was it a case of natural infection (the controls, not inoculated, all died)? Here were 994 pigs and hogs inoculated within a few days of each other, with the same virus, and 946 which later received one full cubic centimeter of a virus (with the Axtell exceptions) selected by myself from a fresh hog in a fresh outbreak, and yet from both inoculations, within three weeks, not one of the 946 ever showed a sign of illness to amount to anything, and none were injured.

The difference of individual susceptibility argument brought forward by the government will no more work in this case than in the Surprise case in 1888, for the only hogs which were sick at all in the whole 994 were those of Mr. Pamblade, at Axtell, who lost 30 out of 73, Mr. Lundren losing none. Mr. Pamblade's hogs were not taken ill until after January 1, 1892, inoculated August 14, 1891, four months, which excludes all possibility of inoculation having been the cause, and he reports that "they did nicely up to about January 1, 1892." So here we have one partial failure in 946 pigs and no injury whatever. If that is not sufficient evidence that the virus did not injure or cause disease in the state farm pigs I can give no other. It is a rank impossibility. Farmers should also bear in mind that the germ of that terrible second swine plague was also injected into those hogs, mixed with the genuine germ in a very weak condition, so weak it would scarcely kill a rabbit.

That experiment proved what we desired, that a hog sick for some weeks and emaciated, showing surely it has been ill some time, is unfit for virus. Were it necessary I could give quite a list of failures due to virus made from pigs not ill as long as that one, for I have tried them in all stages, from those just taken to those sick three or four weeks, and in every case the virus was only reliable and lost in reliability from a pig just taken in a fresh outbreak to one ill longer than a few days; when ill over two weeks the virus from such a hog is useless.

I have also tested virus made from pigs in the earliest days of an outbreak, and then every few days along its course until the last showed signs of illness, and also found that one could not rely on virus from the pigs showing signs of illness after an outbreak had continued much over a week. I look for the cause of this singular phenomena in slowness or fractional, day by day, infection and that the action of the products of the tissues of the body exerts such an influence on the germs when even so short a time in the pig as to render them unreliable or unsuitable for preventive inoculation.

The fact can be absolutely relied upon for sure guidance in obtaining a reliable virus, whether the hypothesis be a correct explanation of the cause of the loss of prophylactic power in the germs or not. If tested on pigs the germs will be found to lose in virulence, also when taken from sick pigs along the course of a protracted outbreak, or if taken from the veins of a sick pig in a mild outbreak. In other words, *the longer a pig is ill, the longer an outbreak continues, the less virulent the germs from such pigs are to pigs themselves in inoculation. Protection becomes null.*

Virulence in Cultures

has

no relation whatever

to

THEIR PROTECTIVE POWER

in

NON-RECURRENT DISEASES.

VIRULENCE IN CULTURES.

Following the teachings of Pasteur and the generally accepted views of the period when I first began work, I also thought that the prophylactic power of the etiological germs in non-recurrent diseases was directly related to their virulence; or in other words, that it was the same product of such germs that produced the disease which also produced the resistant power in the recovered individual that enables it to resist infection or manifest disease on a subsequent exposure to infection in the same disease. The only departure that I made from Pasteur's and others' teaching and example was that the fact that my endeavors to prove that the germ found in pure cultures from diseased pigs *not only seldom had the same degree of malignant virulence in large numbers of much younger pigs used for such experiment; but also, that it was even quite a rare thing to produce fatal results in young pigs by the subcutaneous inoculation of even one cubic centimeter of bouillon cultures.* Let me again say, that in all my work where I wanted pure cultures direct from the diseased hog, I not only used those most recently diseased, but soon learned that in order to be sure of getting such, I could not depend on the farmer's knowing which were recently diseased in the large herds kept in the west, so that I also selected the freshest outbreaks I could find. Following this rule almost invariably I have with equal certainty obtained pure cultures of the swine plague germ from both the heart's blood and spleen of diseased hogs. Until experience had conclusively shown me that this method could be relied on I had invariably had recourse to plate (control) cultures, both on agar-agar and gelatine; but finding I could depend on cultures thus obtained for purity for general purposes I soon left them off as unnecessary, for it must be remembered that I was "man of all work," even to cleaning my floors and feeding my animals and keeping them clean during my first year here, and hence all unnecessary labor had to be carefully avoided. The microscopic appearance of cover-glass specimens from my solid media and fluid cultures, with the general uniform character of the cultures was considered and found to be a sufficiently reliable guide to their purity. The fact that pure cultures can almost always be obtained in this and kindred diseases—septicæmiæ—has been seen by the numerous gentlemen who have worked with me here and is also known to every investigator, for when we have mixed cultures do we not inoculate some very susceptible animal and thus get the pure culture of germ desired directly from the heart's blood of such animal, and do we not do this to avoid the time and trouble of plate cultures?

It must be remembered that during my first year here that I used young pigs, because I had no place to keep rabbits, and they were more available. Again, in obtaining cultures in this way the presence of adventitious germs is almost entirely shut out, and as I was only seeking the swine plague germ, and experience having taught me that they could be obtained pure by following the above given rules, and as good sense told me adventitious germs would certainly be in the lungs, I did not use any other tissues for cultures, and so escaped coming across that government germ of a nondescript plague, and it was only when I had recourse to rabbits entirely that I found the thing in Chicago. Let me here say, that if any one thinks rabbits reliable in this regard they will be most seriously misled. For example, let them take a piece of badly diseased lung from a swine plague hog, and an extra malignant outbreak (and from it also make a number of plate cultures), and rub it up in sterilized water and inject a cubic centimeter of the material under the skin (or into the abdomen, which is surer) of a young pig (two months old) and then one-fourth of that amount under the skin of a rabbit; the rabbit may die of the government germ and the pig of genuine swine plague, *if it dies;* the plate cultures will also lead one astray unless the medium is rather alkaline, for the government germ may not grow, while the genuine one and adventitious germs, if present, will be found in more or less abundant colonies. Again, though they may find the government germ in pure cultures in the heart's blood of the rabbit, which will generally die in from fifteen to twenty-four hours, if they will carefully sterilize the skin at the *locus inoculationis* and open it with due caution they will undoubtedly find that the genuine swine plague is present there in great numbers, but time enough has not elapsed for its general distribution through the blood and tissues, though plate cultures, especially from the spleen of the rabbit, may even then bring out an occasional colony, whereas the wire may not catch one in cultures made from the heart's blood. No one need to take my word for these assertions. They can prove them all if they desire and will have due regard to exactness, and particularly honesty to the facts they see.

The misleading character of the following statement from the government investigators in connection with their second plague germ must be at once apparent to the unprejudiced reader. Rabbits are totally unreliable unless used with all the precautions suggested. In mixtures of germs small animals are never reliable until demonstrated to be so by all possible controls, especially by such in the species in which a given germ causes a specific disease, for there is frequently some adventitious germ that kills the small animal, which on subcutaneous injections with the natural species are harmless. Pulmonary injections are to be condemned as misleading unless controlled by subcutaneous in the same species, or directly indicated as in pleuro-pneumonia, but even then I should prefer spraying suspected cultures gently into the trachea. The passage is as follows:

Of the 49 animals of the same herd 17 were found with collapse and 8 with lobular broncho-pneumonia; more than one-half, therefore, had some defect of the lungs. It might be questioned whether such lesions as those of broncho-pneumonia are not due to the swine plague bacteria, since they closely resemble the lesions found in many swine plague lungs. *This question is effectually disposed of by the inoculation of the lung tissue into rabbits.* (Report 1887, p. 486.)

The position advocated by me is very strongly supported by that taken in the Johns Hopkins laboratory, where they have probably tried as hard as the government people to prove the second germ to be the cause of a distinct plague and thus far failed, so far as we have any statement from that source:

At the meeting of the United States Veterinary-Medical Association, held at Washington, September, 1891, Dr. A. W. Clement made some remarks which bear on the report of that "board of inquiry," because "it has been my (C.'s) opportunity for the last three or four years of doing some work in that line myself, in association with Prof. Welch of Johns Hopkins University. I might say, in parenthesis, that my work in this line has nothing to do with my position as government inspector. As to what connection the organism (Salmon's swine plague germ) has with the lesions described in the reports of the bureau, is a question on which we all might not agree. Nevertheless the swine plague organism does cause trouble. *The trouble in hogs is, as a rule, in our experience, one of mixed infection. We have not had the opportunity of seeing an outbreak of swine plague (Salmon's) pure and simple.* We have found that it is very hard to say when swine plague (Salmon's) is present that hog cholera (Salmon's) is absent, from the fact that swine plague kills [What, rabbits or hogs? B.] in a few hours while hog cholera requires some days. If, then, an animal be killed and presents lesions in the intestines, as are generally supposed to be characteristic of hog cholera, the statement must be very carefully considered before it is made, that hog cholera is not present. We were thrown off our track during the earlier part of our investigations. *We found afterwards that hog cholera did exist in these animals that we thought had swine plague (Salmon's) pure and simple. I would simply say in a general way that form our investigations we have found Dr. Billings is right in certain other matters.*"—(*Journal of Comparative Medicine*, Vol. XII, p. 549.)

Having then found:

First—That the most recently attacked animal from a very fresh outbreak could be relied on to give pure cultures if handled *lege artis;*

Second—That even one cubic centimeter of a fresh bouillon culture was not by any means generally as deadly in subcutaneously inoculated pigs, and seldom so except from the most malignant outbreaks;

Third—Using no other cultures during my first three years here, except just before I left, and having never had a failure on hogs thus inoculated when put in outbreaks;

I naturally came to the conclusion that artificial mitigation was unnecessary and never have resorted to it, but I still thought that virulence was directly connected with the prophylaxis. But I had done more. I felt that I should, if possible, have some means of controlling the virulence, and so, following authority, I selected rabbits and found that a virus, of which one-half cubic centimeter would invariably kill a rabbit in about four days, could be given to six months old pigs in one cubic centimeter doses without danger. As I shall show, I was all wrong. The control of virulence in rabbits with this germ is absolutely valueless as regards pigs. No matter how virulent and malignant the disease may be in pigs, no matter if an outbreak knocks out 100 head of good healthy pigs within a week, or even less, and many do not seem ill more than twenty-four hours, yet an injection from a fresh bouillon culture of one-half cubic centimeter will have no great variation in rabbits, the killing time will vary from three and one-half to four and one-half days. Now carry that same culture in bouillon indefinitely and six months afterwards try it on rabbits and then in pigs intra-abdominally; it will kill the rabbits just as before and may not affect the pigs at all. Subcutaneously it will not. To demonstrate this fully, let me say that I have here a culture from an original culture made in 1888 that has been changed weekly to fresh bouillon ever since. The virus was over a year and a half old when these tables were made. Both are from two different hogs from the same outbreak and were treated exactly alike.

Virus No. 1.

Dose, one-half cubic centimeter.

Generation.	Date of Inoculation.	Killing Time in Rabbits.
1	September 28, 1888	4 days
35	January 5, 1889	4 days
76	October 2, 1889	5 days
78	October 16, 1889	4 days
79	October 23, 1889	4 days
80	October 30, 1889	3½ days
81	November 10, 1889	4 days
82	November 17, 1889	3½ days
83	November 25, 1889	4 days
84	December 2, 1889	4 days
85	December 9, 1889	4 days
90	January 26, 1890	4 days
93	February 22, 1890	3½ days
95	March 9, 1890	2¾ days
98	March 31, 1890	6 days
99	April 6, 1890	4 days
100	April 13, 1890	4 days
101	April 20, 1890	4 days

Virus No. 2.

Dose, one-half cubic centimeter.

Generation.	Date of Inoculation.	Killing Time in Rabbits.
1	September 28, 1888	4 days
77	October 5, 1889	4 days
78	October 13, 1889	3½ days
79	October 16, 1889	4 days
80	October 23, 1889	3½ days
81	October 30, 1889	3½ days
82	November 10, 1889	3½ days
83	November 17, 1889	3½ days
84	November 25, 1889	3½ days
85	December 4, 1889	5 days
86	December 11, 1889	3¾ days
90	January 10, 1890	3¾ days
95	February 22, 1890	3½ days

A recent test of virus No. 1, May 2, 1892, in which a rabbit recived one-half cubic centimeter of a bouillon culture subcutaneously, killed the animal in three days and twenty hours. We then tried a peculiar feeding test in pigs, which is the most reliable of all tests for the swine plague germ, and used this virus, but used mixed cultures made by inoculating the same flasks directly from tested pure cultures of this germ and the government swine plague germ, and at the expiration of four days' developmen tested the cultures in rabbits; result, death, seventeen hours, government germ in heart's blood, swine plague germ local, one colony of the latter grew in six plates made from spleen.

Experiment: Three six months' healthy and never-exposed hogs fed 2,500 cubic centimeters of bouillon culture for four successive days with this mixed culture; ill a few days, off feed, one had a slight diarrhea; *all are alive and well now.* In this experiment I hoped to show that the real germ would give entrance into the blood of the government germ by causing lesions in the intestines.

Two hogs, same litter, fed with same quantity of a pure bouillon culture of the government germ at the same time. No effect! This accords with the statement of the government that feeding experiments with that germ are futile, which makes it necessary for them to explain the intestinal lesions which they claim are caused by that germ alone, and not by the genuine germ.

The above results show the futility of any general dependence on results in small animals in relation to those in which a given germ causes a specific disease. *It is only when proven of specific value*, as in glanders in male guinea pigs, that any reliance can be placed on tests in small animals. Otherwise thier value is only that of a supplementary laboratory aid—*a cheap living isolating medium*, if I may use the term.

As is well known, self-respect caused me to leave Nebraska in 1889, on account of the persistent intrigues going on against me at the instigation of the chief of the Bureau of Animal Industry. I left here June 20. I had done no inoculation since the previous winter, but before leaving, at the urgent request of Mr. Allen, of the Standard Cattle Company, of Ames, Neb., I inoculated fifty-nine pigs, and from there went to Gibbon, Neb., and inoculated 614, some of them being only two weeks old. Those at Gibbon were at once dropped, and ran around as before; *none were injured in any way.* Mr. Allen's were penned in a small pen, and he reports the loss of four. I did not then know the danger of close penning of inoculated hogs, or that in it and not in the inoculation was to be sought the cause of some pigs being stunted in a few cases, which led me to some erroneous conclusions that have been taken advantage of by the government, as follows, in its Bulletin No. 8:

Being requested to make this experiment by the state veterinarian, in order that the method might be adopted in the field, if successful, by the State Live Stock Sanitary Commission, he replied in a long article from which the following quotation is made:

"I leave it to every practical farmer in Nebraska whether he considers another test necessary to show that prevention by inoculation can be done. Then why is it not practical as well as practicable?

"First—Because it stunts the hogs in their growth.

"Second—Because the method used consists of a virus which contains the germs or specific cause of hog cholera.

"(So long as we use such a virus as that, so long will it be for every hog thus inoculated to infect the earth or pens where it is placed, and hence make pestiferous centers where none may previously have existed. As the earth is the natural abode of the germs of this disease, it is self-evident they would again acquire their natural virulence in course of time.)

"We have not been engaged to spread the disease but to prevent it.

"These two circumstances were doubtless unknown to the state veterinarian of Nebraska when he suddenly displayed such extraordinary interest in the welfare of the swine breeders of the state. They show the utter folly of continuing this line of experimentation and the test demanded by him."—*Nebraska State Journal*, October 9, 1887.

The above quotation shows that thus early in his investigations the investigator recognized three conclusions as the result of his experiments: first, that inoculation stunted the hogs; secondly, that it spread the disease; and, thirdly, that the method is not practical, and that it is utter folly to continue this line of experimentation.

One of the reasons I would not make the test desired was the same that caused me to refuse all connection with it at Ottawa, Ill. I did not dare to put my reputation so squarely into the hands of those whom every one here knows would stop at nothing to ruin me. Regarding the dangers and injuries I was mistaken. The hogs that led me to the idea that inoculation stunted them were confined in pens six feet square at the state farm for months. The sale records of the state farm will show that afterwards, when let out free, they overcame it and when killed weighed over 500 pounds each.

This is not the first time the Agricultural Department has brought up this matter, but it has carefully withheld the direct proof which I gave, in 1890 that it was the confinement and not the inoculation which did all the injury. In no sense of the word have these persons displayed any knowledge of the meaning of the word honor, and the secretary of agriculture is partner with them in this dastardly and dishonorable course. "Evil communications [seem to] corrupt good manners" in Washington as well as elsewhere.

In my paper on "Preventive Inoculation," read before the American Medical Association at Nashville, Tenn., May, 1890, I said:

Virus No. 5.

Generation.	Date of Inoculation.	Killing Time in Rabbits.
1	December 19, 1889	2½ days
2	December 24, 1889	2½ days
3	January 1, 1890	4 days
4	January 10, 1890	4 days
5	January 19, 1890	4 days
6	January 26, 1890	4 days
8	February 22, 1890	3½ days
13	March 31, 1890	4½ days

From the first generation of this virus No. 5, seven healthy hogs received one cubic centimeter in the inside of the thigh subcutaneously; two of these died, which were closely confined, within ten days; the others were at first loosely confined, and showed no ill effects; but after the fifteenth day they were changed to the closest confinement possible, and all died between the thirtieth and thirty-fifth days; while five others, which were given plenty of room to roam about, and which received the same dose, all lived. This experiment was made to demonstrate, if possible, a certain well-known fact of practical experience.

It has always been told me that if such hogs were shipped, the death-rate would be checked, or even stopped, while on the cars; and it has also become more or less current among farmers that to put sick hogs on a common wagon and rattle them over frozen and rough ground was a good thing to do. In some few cases evil results have followed inoculation. Some few hogs have either died after a prolonged illness or become somewhat stunted; while others inoculated with the same dose of the same virus have shown no ill effect. It therefore becomes an interesting question to discover why this should occur in one case and not in a great many others.

After very careful inquiry it was discovered that, where ill effects followed inoculation, the animals were kept very closely confined, and had not room enough to move around, and that this was the cause; or, in other words, insufficient movement to stimulate the circulation. A very striking example of this occurred in two bunches of hogs inoculated at the same time for a gentleman who is a personal friend and a great advocate of inoculation [Mr. Lee, of Gibbon, Neb., quoted elsewhere]. At his house were twenty-five hogs in a small pen, while in a field, running with his cattle, were about one hundred and twenty-five others. Those in the small pen had a prolonged and very severe attack of swine plague in consequence of the inoculation, though none died; while the one hundred and twenty-five in the field showed no ill effects whatever.

Exactly the same thing occurs in typhoid fever in man. The patient has recovered from his typhoid, and the physician congratulates him; the next day he is cyanotic, rapid breathing is present, and pneumonia develops; owing to stagnation of the circulation, and the reflow, or pressure, of the blood to the point of least resistance, which is the lungs.

Would not the spirit of common fairness which all gentlemen try to observe towards each other demand that Secretary Rusk should have at least noticed these remarks of mine, even though he might say that he did not accept the explanation. "Fairness," however, seems to be absolutely an unknown quality in the minds of my opponents.

The virus used in the hogs at the Standard Cattle Company and at

Gibbon in 1889 was a later generation, exchanged weekly, of the same used at Surprise, Neb., in 1888, and then used in the first generation, and which in every case (but in the herds diseased when inoculated) is known to have given immunity not only to herd exposure, but twenty of Mr. Ashburn's hogs (at Gibbon) inoculated at that time, 1888, were inoculated with two cubic centimeters of one of the most malignant cultures I ever had, and the thirty inoculated at the state farm, done when Mr. Ashburn's were, also received the same dose with seventeen uninoculated controls, all of the latter were sick, and fifteen died. *Not one of the others ever lost a meal.* The government knows all this, but fails to mention it.

Now, here was a virus that was known to have given protection. It was tested on rabbits and killed in the usual time. Its virulence was known not to be dangerous and proved to be so in the 614 inoculated at Gibbon, many of which were sucking pigs two weeks old, and I remember I cautioned the owner of my lack of experience with such small animals, but he said, "go ahead, they will die any how, and if there is anything in it I want it." No experimenter ever had better supporters than those farmers around Gibbon, Neb. I was leaving Nebraska, and so on my return from inoculating the pigs at Ames and Gibbon I wanted to be dead sure of my virus, and at once fed four three-months-old pigs 300 grammes a day of a bouillion culture of the same material for five days. Suffice it to say, they died of swine plague before I left.

Now, what more could a man ask according to the mitigated virulence theory of protection by inoculation. Here were all the necessary conditions guaranteed *lege artis:*

First—A properly selected virus that had given protection in severe tests.

Second—That had the same virulence in June that it had in October by rabbit controls.

Third—A safe relation between a fatal dose in rabbits and a given dose in hogs.

What more could a man ask if the virulence theory of protection was correct?

It was and is all wrong!

The Ames pigs did not stand a subsequent test, *though four died from swine plague due to the inoculation.*

In Bulletin No. 8 occurs the following:

The attempt to protect hogs by inoculation when fed in distilleries was a failure.

Our tests in distilleries were a success, and a marked one if the whole truth was known, but that is just what the Hon. J. M. Rusk does not desire to be known. After making the arrangements at Chicago which enabled me to carry on these investigations, much to the disappointment of that friend of the farmer, Hon. J. M. Rusk, I decided to try a test in the distilleries, and purchased in Nebraska thirty hogs inoculated in 1889 from Mr. Bassett, of Gibbon, and twenty from Mr. Walker, Surprise, inoculated with the same virus, when fresh, in 1888, but by mistake two uninoculated hogs were put in by Mr. Walker; *these two died*, the others not only did not, but were carted over Illinois and Indiana and kept in outbreaks until so large they could not be moved any longer, and the whole eighteen, as far as I know, were afterwards marketed. I saw several of them when they weighed over 500 pounds, notably two Berkshires which Mr. A. M. Caldwell, of New Holland, had. Of the Bassett hogs fifteen of the thirty died. We purchased a dozen sick hogs and put them with this lot.

What honest man would call that a failure? There must have been some reason for such a singular difference in deportment. The chief of the Bureau of Animal Industry has made the dastardly assertion that the eighteen hogs from Mr. Walker's were "recovered hogs." That is an out and out lie, no matter who so "informed" him. He knows, but dare not say it, that even had they been, his very assertion is simply the reiteration of the natural proof that inoculation is possible—*for the disease is non-recurrent.* All we have to do is to learn to repeat nature's handiwork. We are doing our best to succeed, while the "dogs in [Mr. Rusk's] manger" will neither do anything themselves nor permit any one else to if they can help it.

A feeding test at the distillery at Terre Haute, Ind., under circumstances so terribly disadvantageous that I will not burden the reader with the full details, with sixty-eight inoculated hogs, made especially for Mr. George A. Seaverns, resulted so miraculously favorably as to lead him to go into that mistaken venture at Davenport, Ia., with glucose slops, because he could not get those in any distillery, they all being controlled by a trust. No such test of preventive inoculation was ever made as this and none need ever be made again. The loss was twelve out of sixty-eight against a square loss of 100 per cent in the farm herd from which we took the sick pigs, and the conditions at the distillery indescribably unfavorable. The hogs were in muck up to their eyes, not a drop of fresh water for mouths, nothing but slop, and no shade; with the July sun pouring down on them so hot that no man could stand its rays very long unshaded. It was constantly over 120° on the ground the whole day during most of the time from 10 A. M. to 6 P. M. while the test lasted, and the distillery building put the hogs in an airless box open only towards the south and the river. Their feed was shut off and the sick hogs put among them and as they died cut up and eaten by the inoculated hogs. If this was not a successful test of inoculation in distilleries I do not know what to call success.

TO RETURN TO OUR HOGS INOCULATED JUNE, 18·9.

Something caused the success in the Walker hogs and the failure in the Bassett lot, both being inoculated with the same virus, one lot in October, 1888, and the other in June, 1889, which caused me to stop and think, for those June hogs were the first I had failed in. At once it came to me that in every other case I had used such a hog for virus as I used to obtain pure culture, viz., *a fresh hog in a fresh outbreak not too malignant*, and from such had always been successful. Was it true, then, that we were all on the wrong scent and that virulence has no relation to the prevention in regard to the products of these germs and that the protection had been lost in the cultures?

When I left Nebraska depending on this natural low degree of virulence theory, and on the supposed safety relation between a given dose in rabbits and a proportional injection in hogs, I thought the problem solved and that all I had to do was to continue this culture indefinitely and use it as a preventive virus. With this idea satisfactorily fixed in my mind we began inoculations from Chicago, which antedated the experiment at the Peoria distilleries, to have our eyes opened by some failures, the first being in the hogs of Mr. A. M. Cadwell, of New Holland, Ill. Seeing, then, that there must be something wrong in that virus, I fed six pigs with it until they were evidently quite sick, stopped the feeding and they finally all recovered, though they emaciated much in the meantime. These were what might be called "recovered" hogs. I then procured five hogs inoculated with a fresh culture from a fresh outbreak and an equal number of "recovered" hogs from a natural outbreak, and put them all in a severe outbreak just beginning. The six animals fed with the old virus all died; the hogs inoculated with the fresh virus and the recovered hogs all lived, and, in fact, were not sick at all. Rabbit tests were again made with the same virus and they died as before, and still do in about four days.

These experiences conclusively demonstrated to me that we were on the wrong track and that a control of virulence either in small or any kind of animals was absolutely useless unless as a guide with regard to the prophylactic power in a culture made of the germs of a non-recurrent disease unless experiment proved them to have value.

The first investigator who experimentally demonstrated the fact, so far as I know, that many pathogenic germs produce a toxic or disease-producing substance and an immunity-producing material, was Chauveau, who published his investigations in the *Archives de Med. Experimentale*, March, 1889. These experiments and experiences of mine came in the latter part of the same year, and conclusively confirm the observations of Chauveau made in relation to bacillus anthracis. In my address made before the American Medical Association at Nashville in May, 1890, I referred to my work as follows:

To discover this fact has cost an immense amount of experimentation and a large amount of money. There is one difficulty about it which cannot be overcome. While the virulence of a culture can accurately be tested on small animals, and the exact relation be established between that virulence and a safe dose in the animals in which a given non-recurrent disease occurs naturally, we have no means of testing the preventive properties of a virus, except the results in inoculated animals of the species in which the disease naturally occurs, and this takes a long time and a great many inoculations.

When I began this work, I thought that a test of virulence, with a proper regulation of the proportional dose, was all that was necessary; now I know that a test of virulence with such proportional control is of no value whatever with reference to the protecting power of a given virus. I know this by an experience in over one thousand five hundred hogs. Either one of the six viruses named above will kill a hog by subcutaneous inoculation in sufficient doses, but more reliably by feeding experiments; yet not a single one of them will protect a hog an iota against the cholera; while a germ virus, without any virulence whatever, even by forced feeding, will protect securely and almost invariably, in single doses of one cubic centimeter.

This assertion is obviously contradictory to all previous experiences, and yet, if one stops to think a moment, supported by the most trustworthy experience which we have.

There is no question that, under certain circumstances, using a natural virus, we must reduce the virulence to a certain degree in order to produce a fatal attack by inoculation; but on the other side, we do not need to have a virus possessing any virulence whatever to produce immunity in a non-recurrent disease.

Vaccination gives the best possible example of the point we desire to call attention to. While in variolation small-pox itself was transmitted, and the variolated person was as much a source of contagion as one having the natural small-pox, the vaccinated individual is absolutely non-dangerous to those coming in relation therewith.

What has been lost?

Certainly something; and the practical evidence conclusively demonstrates that it must have been, or is, that peculiar element of the germ, the contagion, which rendered the variolated person dangerous while the vaccinated one is not.

It is a question of nutrition only, which can be demonstrated by exact experimentation if we only try.

The bovine organism must offer certain nutritive conditions which, in some unknown way, rob the micro-organismal cause of small-pox of its contagious (small-pox) producing qualities, while the preventive one is retained in *optima forma.*

The same thing can be done with the germ of swine plague, but to no such degree of absolute certainty as nature or accident has accomplished in vaccina. While it is the easiest possible matter to feed the germ of swine plague up and down in virulence, to make it extra malignant, or rob it of that quality altogether by changes in the chemical nutrient, it is a most difficult thing to retain the preventive physiological qualities of these germs. Sometimes it can be done for months, and again they are lost in a few generations; but we have no control over these matters except the practical tests. None of the small animals which I have tried can be easily or successfully rendered immune against swine plague. Rabbits and guinea-pigs can be rendered somewhat immune by repeated inoculations of small quantities of virus; but, so far, I have been unable to render them absolutely so. Pigeons vary; but no more artificial immunity can be produced in them than they naturally have, as will be shown later and at another time.

As has been repeatedly mentioned, a test or control of virulence has nothing whatever to do with the protective power of a virus. It simply shows that it is safe to use, and that we can establish the point of safety by experiment.

This shows that we have all been working on an erroneous basis. Prevention has no relation to specific virulence. Others have demonstrated the same fact.

In *The Times and Register*, Philadelphia, of April 12 and 19, were published a series of most interesting experiments by Chauveau, the most eminent and conservative experimenter in France, in which he demonstrates it in connection with bacillus anthracis. Chauveau says* that "Energetic vaccinal (preventive) properties have been discovered in a pathogenic germ (B. anthracis) not only attenuated in its virulence, but systematically deprived of all infectious properties—rendered so neutral and inactive that we are forced to ask ourselves if this transformed microbe had not become a new species.

"Cultures of bacillus anthracis in this condition can then be carried on in the ordinary atmosphere."

Chauveau conclusively shows that bacillus anthracis produces two chemical elements in its bio-physiological development; the one toxic, or specifically disease-producing, the other having exclusively the preventive properties; and that, by cultivation in two to two and one-half pressures of oxygen, the toxic properties may be, or are, lost, while the preventive are retained. Or, to use his own words, "In fact, in my experiments the vaccine property of the transformed bacillus anthracis is so active, and so well survives the loss of all infectious properties, that we seem authorized to consider these two properties as being absolutely independent of each other, and as each belonging to a special product of microbe life."

Another point in evidence of this fact is this: If we take a culture of the germ of swine plague which has experimentally been shown to actively possess both of these preventive and toxic, or disease-producing qualities, by actual experimentation, and freeze it solidly for several days, and then inoculate or feed hogs with it, we will find that, while it has retained its toxic, or disease-producing, properties without mitigation, it has entirely lost its preventive property.

That bacillus anthracis has the power of producing these two essentially different chemical elements has also been well shown by Hueppe and Wood.

Though I sometimes criticise the conclusions of my friend Hueppe, still I think him one of the most competent and reliable patho-bacteriologists living, especially in regard to physiologico-chemical attributes of pathogenic bacteria. In the publication mentioned, Hueppe and Wood describe a saprophytic bacillus absolutely without virulent qualities, which, in every method of artificial cultivation, or under the microscope, bore such close resemblance to B. anthracis that it could not be distinguished from that organism. It is a well-known fact that all previous experimenters had not been able to render mice im-

* The original appeared in *Archives Med. Experimentale*, March, 1889.

mune to anthrax by any system of preventive inoculation; and yet, with this absolutely non-virulent germ, Hueppe and Wood were successful in rendering these most susceptible animals immune to extremely virulent cultures of bacillus anthracis.

Now, it is neither logical nor reasonable to suppose that this was any other micro-organism than baccillus anthracis. It was derived from the earth possessing exactly the attributes described by these observers. In this case the nutritive conditions in the earth had naturally produced exactly the same physiologico chemical conditions in bacillus anthracis which Chauveau has conclusively demonstrated to be possible of production by the cultivation of the same germ under certain degrees of oxygen pressure, and I have been able to do with the swine-plague germ by chemical nutrition, the only difference being that while I have, as it were, "gone it blind," not being a chemist, and accidentally hit on a means of arriving at a certain practical result until I have empirically arrived at a method of obtaining a certain result for an uncertain length of time, Chauveau has found a definite means of obtaining it with reference to the bacillus anthracis.

That the chemical nutrition method is the one by which practical results will eventually be obtained goes without question; but it remains for the chemist alone to really discover and perfect it.

The discovery of Hueppe and Wood regarding bacillus anthracis finds its confirmation in many diseases of extra-organismal origin, and explains that heretofore mysterious condition known as "acclimatization" in diseases of this character, such as yellow fever, southern cattle plague, etc. In these cases the specific germs, in a saprophytic or non-malignant condition, must have gained entrance to the individuals possessing this acclimatization immunity while they possessed this prophylactic power, though not possessing the toxic or disease-producing.

That all these exogenous germs are, or can be, changed to saprophytic, must be self-evident; and that their toxic or disease-producing property is acquired by peculiar nutritive conditions in the soil which they naturally inhabit, seems also equally clear. In fact, their acquisition of disease-producing qualities is dependent upon the prolonged saturation of the ground with the excreta of animal life or the decayed products of animal tissues. The delicacy of the action of these germs in different nutritive media is so little understood, and has enjoyed so little experimentation, especially chemical investigation, that we really know very little about it; yet it is the open field of original research which will eventually lead to success, and from which we can only hope for decidedly practical results.

Why the swine plague virus should lose its virulent properties in cattle, or even when cultivated in sterilized cattle urine, is a question no one can decide at present. Why the germ of the corn fodder disease should not be toxic to animals while still manifesting its presence by specific lesions in green and growing corn, and only become toxic when and after the leaves begin to wither and the chlorophyl suffers chemical changes, are also questions of a nutritive nature, which can only be elucidated by the most exact chemical investigations.

At one time in their existence all these organisms are saprophytic, and again they become pathogenically toxic, all of which is determined by the material they develop in.

This fact that pathogenic germs do produce a toxic or disease-producing material and an immunity or resistance-giving substance has been still more fully confirmed by the later researches of Breiger, Kitisato, and Wassermann in the investigations "*Ueber Immunity and Gift-festigung*" published in the *Zeitschrift für Hygiene*, Vol. 12, p. 137, 1892. They say: "Our experiences with the germs of cholera (Asiatic) and tetanus justify us, we think, in formulating the axiom, *that the toxic and immunity-producing principles are two entirely different things*" (p. 162); and again: "by the repetition of these experiments we found, almost without exception, *the existence of the protective power in the less virulent cultures in contrast to the virulent*" (p. 163).

It will be seen, by referring to my own remarks again, that I said that "*Vaccination gives the best possible example of this condition, that is, that the protective properties of a virus have no relation whatever to its virulence.*" In other words, that virulence is what we desire to get rid of entirely and to strengthen and retain this resisting product in our artificial cultures. I called attention to the fact that the "bovination of small-pox," which is vaccina in the cow, had brought this about to the utmost perfection, as a virus to prevent small-pox, without an intelligent act on the part of man in the first place.

As usual, the authors of the publications of the Agricultural Department find it necessary to contradict this statement, and with their usual reckless inconsistency with regard to previous statements of their own. In this "Bulletin No. 8" they say:

Some breeders have advocated inoculation on the ground that vaccination had been found efficacious in preventing small-pox in the human subject, and that, consequently, inoculation should be an equally reliable preventive of hog cholera. In reaching this conclusion they overlook two very important facts. In the first place there are communicable diseases, such as tuberculosis, from which no immunity can be acquired, either from vaccination, inoculation, or an attack of the disease contracted by ordinary exposure. It is therefore impossible to decide such a question by reasoning from one disease to another. The matter of immunity must be determined by observations with each particular disease. In the second place, the effects of inoculation and vaccination are radically different. *The vaccine virus, as used in the prevention of small-pox, is not the virus of small-pox, but of a different and distinct disease. It produces a mild disease in cattle, and an equally mild disease in people. It never assumes a malignant and fatal character either in cattle or people.* For this reason it can be used with safety. Before vaccination was discovered, however, inoculation with small-pox virus was sometimes used, but its results were uncertain and often fatal.

Just how reliable the following quotation is I cannot say, but can see no reason for doubting it. If true it puts a clincher on the government's statements as to variola and vaccina being different diseases. The *Medical Record*, August 6, 1892, copying from the *British Medical Journal* says: "Dr. T. W. Heim has successfully transformed small-pox into cow-pox by inoculating a calf with variolous lymph" (p. 164).

If swine plague is not a "non-recurrent" disease then nature is a liar. Tuberculosis is not a "non-recurrent" disease. Such reasoning is pure humbug. Let us quote one passage again to be sure of our ground: "The vaccine virus as used in the prevention of small-pox is not the virus of small-pox, but of an entirely different disease." Let us see what these same authors said on that very subject in earlier reports:

In the following chapter the results of our experiments with the vaccine (Pasteur's) will be given in detail, and from them it will be seen *that it does not prevent swine plague for the simple reason that the vaccine of one disease cannot protect against another.* (1885, p. 187.)

Again:

The above experiments with Pasteur's vaccine, do not, in our opinion, therefore disprove the protective power of Pasteur's vaccine over rouget, *but simply show that the vaccine for one disease will not protect against another.* (1885, p. 194.)

For the same reasons preventive measures applicable to one cannot be applied to another if there are differences in the microbes that cause the disease. (1888, p. 193.)

What more evidence do we need that "consistency is (not) a jewel" known to Mr. Rusk or his advisers. On the other hand, we have some very competent observers who have come to the same conclusions that we have, such as:

> *Each of the tested varieties of serum protected only against the species of bacteria which the animals were made immune by from which the serum was derived. It had no value against infection due to other forms of bacteria.* (Klemperer: *Ueber die Heilung von Infectious Krankheiten.* Berlin, *Klin. Woch.*, 1892, No. 18, p. 422.)

> *It was demonstrated that the concentrated fluid possessed exactly the properties of the original fluid: it is not toxic, but produces immunity.* (Klemperer: *Ueber die Heilung von Infectious Krankheiten.* Berlin, *Klin. Woch.*, 1892, p. 423.)

> *Probably different substances are produced by bacteria in their development, of which some in proper quantity are suitable and can be used to produce immunity, while the others are only injurious.* (Roux on Immunity, International Congress of Hygiene. *Deutsche Med. Woch.*, No. 16, p. 362, 1892.)

This evidence is sufficient to show that we are on the right track and that a number of experienced investigators in Europe have come to the same conclusions and are working in the same direction.

THE PROPHYLACTIC POWER CANNOT BE MAINTAINED IN CULTURES OF THE SWINE PLAGUE GERM FOR ANY LENGTH OF TIME.

It has been said that the bovination of small-pox, that is, the transmission originally of small-pox to milk cows and then from cow to cow by the milkers, eventually produced vaccina, a variola absolutely without virulence,—for a vaccinated child is not a source of danger among unvaccinated children,—but which had the protective power to fully as reliable a degree as the small-pox itself gives to those who may recover from it on a second exposure to natural infection. Man has nothing to do, intelligently, in bringing about this great blessing. He has only carried on the work which nature did, using the milkers as mere passive tools. Thus far we have been absolutely unable to repeat this lesson with the cultures of the germs of non-recurrent disease for any definite period in our laboratories. The nutrition we offer such germs in our laboratories and our methods of treatment are so unnatural that they soon cease to produce this immunity or resistance-producing element, whatever it may be. As to the toxic or disease-producing element of the germ, it seems as if we could keep that up almost indefinitely; at least for some species of small animals, though, as I have shown, my culture of the swine plague germ from 1888, while it still retains its primal virulence in rabbits, possesses little or none for hogs, showing the fallacy of small animal tests as a reliance for effects in an original species of animals in which a disease occurs as a natural infection.

As regards the germ of swine plague, I know that it is a faculatative parasite; that its native home is in certain kinds of soil under certain climatic conditions, and that the porcine organism is really an unnatural habitat for it, though it can be parasitic therein and produce most damaging results, as is too well known. I think, that could we find the germ in a non-virulent condition for swine, direct in the soil, that we should find that it possessed the desired prevention-producing properties in *optima forma.* To be sure this is an *a priori* hypothesis, but we have much evidence in favor of it. In the first place, let me recall Hueppe and Wood's experience with that germ which they obtained from the soil that so resembled bacillus anthracis that they could not distinguish the two germs by any means of laboratory differentiation. *This germ had no effect even when inoculated in small animals most susceptible to anthrax, and yet it rendered them immune to the action of the most virulent cultures of bacillus anthracis.* Will any one assert that that germ was not bacillus anthracis in a non-malignant state? Now, bacillus anthracis, like the germ of swine plague, is a faculatative parasite; its natural home is the earth of certain localities, many of which have become so pregnant with death to cattle and other susceptible animals that they are known as "anthrax centers." We also know that the germ which produces one disease does not produce any chemical product that can protect against another. Such a thing is too absolutely unnatural to be worthy of consideration. Though we have vainly sought and at times thought that we have found a combative destructiveness in the products of two germs introduced into the same organisms, one disease-producing and the other not, because such things can be seen in our cultures, yet no practical results have been obtained in that direction thus far, though such may be found in some cases in the future. This is an entirely different question, however. There is one which apparently offers far more sure and practical results. Whether one of the two kinds of products now known to be produced by pathogenic germs or another, we do know that, like all other living organisms, germs do produce a something which, when accumulated in sufficient quantities in fluid cultures, is death to themselves. This promises the possibility of a cure. That the nutrition is not used up, as has often been assumed, is shown by the fact that other species of germs introduced into the same bouillon, after the death of the first variety, has been proven by suitable tests, develop for a time as well as in fresh media to which no other germs have been added.

As further evidence going to show that the germs of swine plague have their full protective power in the earth is the fact that the longer they remain in the hog the less their protective power, hence the rule: *virus must be taken from a fresh outbreak, and the most recently diseased hog in the outbreak, in order to obtain any positive degree of success.* Another very singular thing is, that virus made from an unusually malignant outbreak, one in which all the hogs are sick at once and when nearly all die about as quickly, has little or no protective power. This is shown by the practical though seldom experience of a hog having the disease twice. Careful questioning of the most observant hog-raiser will bring out the fact that such hogs were one or two left out of such outbreaks as above described, and the same kind of questioning will also bring out the equally valuable fact that the less acutely malignant the outbreak, the greater the number of recoveries, the slower the course of the disease, the more absolute is the non-recurrent condition produced; hence, the need of a second inoculation.

Now, then, these statements being facts, until we know a better way, or can more closely approach the nutrition offered by the natural soil where the germs live, there is but one way open to us, and that is to obtain our virus as close to its natural home as possible, and *that is from a fresh attacked hog and a fresh outbreak,* and one not too malignant. In my directions I said last summer, "one not killing over 50 per cent." That is not quite correct, because in the end the fatality in slow progressing outbreaks, from secondary complications, may be quite large, and more than that, *I should have said not to take virus from an acutely malignant outbreak such as mentioned above, which gives little or no protection, but is too virulent to use.*

The chief of the Bureau of Animal Industry has said:

> The results already obtained demonstrate the danger of spreading the disease by inoculation and *particularly by the method used and recommended by Billings.—(London Live Stock Journal.)*

A Dr. von Mansfelde, of Ashland, Neb., who dares not meet me in open discussion, also wrote privately to a gentleman regarding my method, that it was unscientific, and as the method which he recommends is that of the authorities I will quote a few lines:

As to Billings' inoculation, he is teaching the farmers a cheap method without much expense for apparatus [Is not that a most desirable thing to attempt? B.] Allow me to say that I have suggested a far more cheap and inexpensive method.

The exposure of healthy hogs in outbreaks is the method recommended by this M. D., who neither advises the selection of the outbreak, nor seems to know that in all such exposures the amount of germs taken into the exposed organism is unlimited. The losses of the country from swine plague show what would be the result of following such nonsensical proposition. This egotistical follower of Esculapius goes on to tell his correspondent "What is a modified virus? *Well it is not gained by transmission through animals only.* The application of higher or lower temperature, for a longer or shorter time, subjection to such chemical agents as carbolic acid or bichromate of potash; acidity or alkalinity of culture liquids; the abstraction or addition of oxygen, and other gases; the use of blood of animals, or its serum, refractory, or made refractory artificially, and other means simple or complex in their character and application, enter into the formation of modified virus." So "a modified virus is not gained by transmission through animals only." The learned physician does not seem to know that the only constantly reliable virus is one freshly taken from the animal—vaccina. He would parade his great knowledge of the methods by which investigators have attempted to produce modified viruses, and as a scientific fact have done so, but seems to be entirely ignorant of the fact that every one of these methods have utterly failed, and are not to be relied on when put to the praccal test, unless we except Pasteur's anti-rabies virus, which is a therapeutic and not a prophylactic remedy, and of which we really know less than we do of any other material we use in investigating, because we do not know the germ.

Notwithstanding the fact that the Department of Agriculture recommended, or said, that the "strongest or freshest virus could be injected with immunity," in 1888, and in 1891 contradicted itself and practiced heat-mitigation for "more than 200 days," still I must say that the very closest study of their reports fails to show me one single case where a fresh unattenuated virus has been once used in their attempts. *One honest experiment of two inoculations made with virus taken from an outbreak of surely not over one week's duration, and if possible from the last pig taken sick, will invariably give protection if used as quickly as a fair development takes place in the bouillon.* The government says this method is dangerous and unreliable. The state of Nebraska says it is not, by retaining me in position. We have tried it and never failed, where all the conditions to success have been fulfilled. We cannot always meet them. In no case has it injured an animal, all assertions to the contrary. The secretary of agriculture says it is a failure, but it has never been tried by his men and so he knows nothing about it. The individual from Ashland, Neb., infers it is not scientific because it is not an artificial method, which clearly shows he does not know what science is.

Science is the endeavor to discover and follow nature's methods of doing things and to apply them to the needs of man.

If we discover the natural method and can exactly reproduce it, we cannot but succeed. Failure is impossible. The closer we keep to a natural method the surer we are of success. The more we depart from it the more liable we are to failure. It is self-evident that we cannot always adhere so close to nature.

Let us apply this logic to inoculation as a preventive against swine plague from the etiological point of view here presented. There can be no dispute that swine plague is a "non-recurrent disease." Hence, if the same method by which nature produces "non-recurrence" can be followed, we must also produce this "non-recurrent" condition in inoculated swine. We cannot find the germ in the non-virulent soil at present and inoculate directly into hogs, or even the mildly virulent one which gives the greatest degree of non-recurrence to protected hogs, so we do the next best thing possible; we use the germ from the least virulent outbreaks, as close to its earthy home as possible; that is, we take it from the hog just as quickly as he gives evidence that he is infected; and to assure ourselves of this point to the best possible degree, we take the animal last taken sick in an outbreak of the fewest possible days duration. We find by several hundred severe farm tests that this method only has given success. Hence, it is as near scientific as we can at present make it, for it is the nearest approach to natural infection possible, because the germs have been in other conditions than those of their natural earthy home, the shortest time possible. We need go no farther than the experiments made by the government investigators to find that all and every artificial treatment of the germ has robbed it of this desired protection-producing power. They admit total failure and so we need not discuss those methods.

It must be repeated that no one unacquainted with the facts can have any idea of the vast numbers of experiments that have been made before we have arrived at the knowledge that success was only obtained in one way. In fact, scientific success is always arrived at by seeking the causes of failure and eliminating them one by one as we proceed. Not only did we have to test hogs from every kind of an outbreak, but at every stage in the outbreaks and every stage of the disease in individual animals, in order to find out within what narrow limits we are at present obliged to work in order to succeed. But more than that. Experience soon taught us that the limits were even narrower than we supposed. Having found out that a virus made from the most recently attacked hog in an outbreak of not over a week's duration and of not too acutely virulent type would give success, we had to learn, *by failures also, that only the first generation of a bouillon culture, and that generation never over a week old, could be securely depended on to produce reliable resistance to natural infection.*

We found that a second transfer to fresh bouillon was not anywheres as near reliable as the first; that a third had but little value, and that a fourth was useless; that is, weekly transfers to fresh bouillon, though the virulence, as tested in rabbits, remained standard all the time. So, at present, he who would succeed can only do so by using a virus made from the most recently attacked hog in a very fresh outbreak, and then twice inoculate his healthy hogs, if possible, when the culture is three or four days old, though, unless frozen, it can be relied on up to the seventh day of development. While freezing seems to have no effect on the virulence of a culture of this germ it absolutely ruins their production of the prophylactic principle.

It has been said that a virus made from a hog sick over a week is unreliable, that the nutrition offered the germs by the porcine organism even in that short time nullifies their prophylactic power. We do not know why this is. It is simply a bare fact supported by numerous failures. Of course such a statement is not to be taken as a cast-iron law; the time will naturally vary somewhat. No more do we know why a virus made from the occasionally acutely malignant outbreaks gives little or no protection. We know the fact again by failures, and the more striking fact, that it is in the recovered hogs from such outbreaks that we find those cases of second infection which are so very rare as to be the proof of the rule of the non-recurrent character of the disease. In the same line comes another striking experience, which is, that a virus made from an inoculated hog that does not withstand exposure to natural infection will not produce immunity, while one made from a freshly attacked hog in the earliest stages of the same outbreak will give it. In this consideration I cannot give all the details of the various failures and controls which have demonstrated these points so conclusively that they may be looked upon as reliable. I am now simply giving results and the causes thereof, as far as I know them, for the benefit of the public and other investigators, if there are any in this country who are unbiased enough to be willing to profit by an experience which now extends over six years of continuous study of this one question.

THE NEBRASKA METHOD THE ONLY ONE BY WHICH TO BEGIN INOCULATION.

The farmers of the country who may read this, and particularly the editors of the agricultural press, should firmly impress two things in their minds:

First—*That swine plague is a non-recurrent disease.*

Second—*This being so, whether another and better method for preventive inoculation may be found, that the Nebraska method is not only the nearest possible approach to that by which nature produces the non-recurrent condition in swine, but that until inoculation has proven to be possible and practical by this method, there is no use or indication for trying for another and better one.*

This should be self-evident to every reader who has any intelligence. I once had great confidence that some method would be found by which the chemical product that produces this non-recurrent condition could be isolated, and for that reason, and in the hope that its promises would lead to the earlier establishment of a suitable laboratory in Nebraska and a competent corps of works once objected, as noted by the government, in going on too fast with the present method, because the chemical method would entirely remove the only rational objection to the present method, the undoubted fact that we do resow the ground with the germ of swine plague; but it is well known that we have a most strict instruction, *that no hogs will be inoculated save on farms where the disease has previously prevailed.* We shall allude to this again. When, however, I had more experience and found out the absolute impossibility of so feeding these germs in our artificial cultures as to reliably insure their producing the resistant principle in any culture but the first generation direct from the properly selected sick hog, and when I considered the immense amount of such chemical product which would be required to fulfill the demand of the farmers, I at once saw that could we even isolate it and demonstrate that it would produce immunity, it would be next to impossible to produce enough to meet the public demand. Hence, I truly believe that no other method of preventive inoculation will ever be a practical possibility.

The investigators of the Agricultural Department have no other way open to them than first to prove preventive inoculation exactly by the "Billings method." Having done that successfully they may be warranted in endeavoring, as we are, to improve on it; but until they have repeated the natural method all attempts to find a chemical method are a waste of the public funds and unwarranted. As in 1883, they are deceiving the public with promises that they cannot practically fulfill.

INOCULATION

NOT

DANGEROUS OR INJURIOUS

WHEN

DONE AS DIRECTED.

INOCULATION NOT DANGEROUS.

Notwithstanding the fact of the complete failure of the government investigators to make even one successful inoculation, because we have succeeded, as we can demonstrate, in every case where the method which we now state will succeed has been practiced, in order to decry our success in those cases, which they cannot deny or nullify, they now shout the terrible cry of danger, and not only do they publish cases so that the inference which would be drawn by farmers is that inoculation killed the hogs in such herds, but they refuse to publish testimony which has been at their disposal, which positively demonstrates that inoculation could not have killed the hogs in the herds in question, but, as has been shown, do publish, or fail to call attention to, contradictions in their own published testimonials from hearsay correspondents who say that hogs died in herds of men who never inoculated, and even in herds that they themselves say "none" died, as in Mr. F. W. Luddon's case at Surprise.

Another case of the same nature is the following:

FAILURE OR INJURY AT TONICA, ILL., IN 1891.

It may also be stated that early in October, 1891, Mr. James Richey of Tonica, La Salle county, Ill., obtained virus from the Chicago establishment organized by Dr. Billings. Mr. Richey inoculated ninety animals, which in his words were "a first-class lot of young healthy hogs." Nine days after inoculation they commenced to die * * * The loss in this case was over 97 per cent.

That statement was printed notwithstanding the fact that I had publicly contradicted it and shown it was impossible. The fact is, I sent the virus from here to Mr. George A. Seaverns, of Chicago, who sent on same day and date the same virus to this man Richey and to R. C. Pointer, Pennington Point, Ill., who used it on same day Richey did, viz., October 9, 1891, some of the pigs being but ten days old. As Mr. Pointer had inoculated several times previous to this and had some interesting experience I append his letter to me on the subject with reference to the pigs inoculated October 9, as well as others:

PENNINGTON POINT, ILL., March 5, 1892.

Mr. Frank S. Billings.

DEAR SIR: I at last acknowledge the receipt of your favor which I received some time ago. In answer to it will say that the hogs I inoculated did well; the spring pigs that were farrowed in March, April, and May I sold in February out of the cattle yard and they averaged 240 pounds; would have been heavier had they been fed in the usual way. I got $4.55 per cwt. The small pigs I inoculated were small for certain; some were only ten days old. After they were inoculated some thirty days, the cholera broke out in my neighbor's hogs only about fifty rods from my hog lot, and my pigs went to their pens and were in their pens with the sick hogs every day during the outbreak and are doing well. I sold six small pigs, one to B. D. Herndon, who took him home and turned him in with his hogs, when they were dying at the rate of four to six per day; he went through all right, and is still all right. Sold one to Ed. Bell, of Bardolph: said pig went through a bad outbreak of cholera, and is still all right. I gave my married daughter a young sow last spring for brood purposes, and she went through an unusually bad outbreak and was hearty all the time and still is, and never showed a symptom of cholera. When I began to inoculate, my neighbors made fun of me at first, but they sold their hogs for $2 to $3 per cwt. when they weighed from 100 to 200 pounds, while I got $4.55 for mine, so the laugh is turned to my side. Will say that if that offer of yours to instruct farmers to make their own virus includes me, that I will surely go to Lincoln and inform myself all I can, as I expect to inoculate all my hogs as long as it works for me as it has the last two years. Wishing you success.

Yours truly, R. C. POINTER.

The virus sent to Richey & Pointer was in the third generation and was only sent to Mr. Seaverns, so no complaint would be made that it could not be got. Subsequently I sent Mr. Pointer a second lot of absolutely fresh virus, which he used on the same stock without injury, as his letter shows, though he does not mention this second inoculation. I may state that while at Ottawa several farmers who pretended to know did tell me that "Richey's hogs were sick when inoculated," which statement is corroborated by the facts. That the virus did not hurt them is further proven from the fact as shown by my own records, that the same virus was used fresh in Nebraska on 399 pigs and hogs from one month to one year old, belonging to three different farmers, and not one injured. The government could have been informed had it had any desire to be honest in this matter. In conclusion let me say that we are going ahead this year as usual, that every endeavor will be used to induce farmers to receive instructions and to force them to make their own virus and do their own inoculation, and we shall win our way out in spite of all and every assertion to the contrary. The government's opposition has fallen as dead as its political tools should lie, so far as Nebraska is concerned.

As further evidence of Mr. Pointer's success I have the following letter of very late date:

INDUSTRY, ILL., July 13, 1892.

Dr. Billings, Lincoln, Neb.

DEAR SIR: I have not received any circulars or papers of any kind from Secretary Rusk. I should have answered just as I did yours. I shall want to inoculate my pigs in about four or six weeks. Will send to you for virus. Will be to see you some time this fall, as it is impossible to leave my work at this time.

Yours, etc., R. C. POINTER.

As further evidence in the same direction, and not coming from myself, for the "prominent farmer" alluded to is the Hon. Samuel C. Bassett, of Gibbon, Neb., secretary of the State Dairymen's Association, and one of those public-spirited farmers without whose aid I could never have succeeded as well as I have, and one who has been most decidedly interested in my work from my advent in Nebraska, I append the following editorial from the *Nebraska State Journal:*

INOCULATION NOT DANGEROUS.

One of the most prominent farmers in central Nebraska calls the attention of *The Journal* to an article in *Rural Life*, published at Waterloo, Ia. Mr. C. L. Gabrielson, secretary of the Iowa Dairy Association, writes as follows in regard to inoculation for the swine plague:

"'Our state never suffered more from hog cholera than during last fall and this winter. It is high time the state took some steps looking toward relief. We think extermination must be resorted to at a general expense.'

"So writes Director Wilson, and we ask: What has the Iowa experiment station done to discover or follow up the claims of others in this line of research?

"We listened to Dr. Billings' address before the Nebraska dairymen, during their recent meeting in Norfolk, Neb., and we have come home with the idea that Iowa could well afford to send a few bright young men to the laboratory of the Nebraska university to be instructed in the secret or art of preparing the virus and experiment on hogs in our own state. There are good men and true over in that state who believe in the Billings' system of inoculation as much as they believe in the existence of Nebraska.

"If the Iowa experiment station folks are dancing to the 'Bureau of Animal Industry' music in its quarrel with Dr. Billings, then so much the worse for Iowa experiment station. Have they investigated the Billings inoculation theory with his cultures? is a fair question to ask. If they have tried the Billings method the results of their tests have never been published that we know of.

"It is to be regretted that of the scientists it cannot be said: 'Behold how good and pleasant it is for brethren to dwell together in unity.' The opposition of government officers to the Billings discovery is to be regretted unless they can fairly demonstrate that it is useless or harmful and liable to spread rather than check the disease."

Our friend goes on to say: "Mr. Gabrielson also incorporates in his article Dr. Billings' method of making virus, which has already been published in *The Journal.* Persons who are interested in the public discussion in regard to inoculation will readily call to mind the Associated Press dispatch from Washington published December 11 last, wherein Dr. Salmon stated that statistics from Nebraska showed that losses from hog cholera had been unusually heavy in the fall of 1891, and also that there had been wholesale inoculation of hogs by the Billings method in that state during the summer and fall of 1891, and evidently deliberately intending to convey the meaning that these heavy losses were the direct result of the inoculations. In view of the effort of Dr. Salmon to show that inoculation is a dangerous procedure and likely to cause heavy losses from cholera—in other words, to perpetuate and spread the disease—the above quotation by Mr. Gabrielson from Director Wilson, of the Iowa experiment station, is of peculiar interest and worth repeating.

"Director Wilson writes: 'Our state (Iowa) never suffered more from hog cholera than during last fall and this winter. It is high time that the state took some steps looking toward relief. We think extermination must be resorted to at general expense.' No one, not even Dr. Salmon himself, can claim that the losses in Iowa were due to inoculation, and yet the same state of affairs, as regards losses from cholera, existed in Iowa as was said to exist in Nebraska. As regards the losses from cholera the past year in Nebraska, in the opinion of the writer they have been greatly overestimated. This opinion is based upon travels the past fall and winter in fifteen of the most populous counties of the state and in attendance at farmers' institutes and like meetings for the discussion of matters relating to agriculture, and from information gained at such meetings it does not appear that losses from cholera have been anything like as heavy in these counties as in previous years. It is also true that in none of these travels or at none of the meetings mentioned has the writer ever heard it charged, either directly or indirectly, that inoculation was the cause of the spread of the disease, or that the disease or losses therefrom could be traced to inoculated hogs. Every intelligent farmer knows that hog cholera flourished before inoculation was known and continues to flourish, but not because of inoculation."

Of all the men who have had hogs inoculated, Mr. Hess of Surprise, Neb., is the only one who positively asserts that inoculation did it, and this he does because he said so before.

In relation to Mr. Gabrielson's remarks as to the relation of the Iowa experiment station to inoculation the following letter and comments may prove interesting to my Iowa readers:

VETERINARY DEPARTMENT,
IOWA AGRICULTURAL COLLEGE, Jan. 25, 1892.

DEAR SIR: Your letter came to hand in due time. In regard to his (Billings') inoculations I will say that I have no doubt but what it is a success in many cases; neither do I doubt but what in some cases it will produce fatal cholera, thus originating an outbreak. I do not believe that farmers themselves are the proper ones to make the inoculation. I have several experiments to make with virus of different kinds as soon as opportunity offers.

Very truly yours, W. B. NILES,
Professor of Surgery and Obstetrics.

The above letter is in my hands in the original. Certainly since January 25, 1892, time enough has elapsed for tests of virus already prepared. Let me say this, as it is admitted that I do succeed "in many cases," the first and only honest way to proceed, on the part of any one else, is to proceed as I have so fully directed and by the methods which have brought on the success which has been so far achieved. The fact is, not one of the opponents of inoculation has as yet, so far as I know, correctly and honestly repeated the methods by which inoculation has succeeded in Nebraska and elsewhere. They have all been most diligent in using the methods of artificial attenuation, which have invariably failed in my hands and therefore have been absolutely condemned. Less jealousy of me and more honesty toward the public would be more becoming in persons who are as much in the public service as I am.

THE MALICIOUS DISHONESTY OF THE DEPARTMENT OF AGRICULTURE SHOWN BY ITS COMPARISON OF LOSSES IN INOCULATED HOGS WITH THE PREVALENCE OF THE DISEASE IN THE BALANCE OF NEBRASKA IN 1891.

If we deduct from the total number inoculated, as given by Dr. Billings' statement (2,952), the number contained in the herds that were said to be diseased when inoculated (394), we have remaining 2,558 as the number inoculated which had not previously been exposed. Among these it is admitted that the loss from inoculation and exposure amounted to 198, or 7¾ per cent. This is nearly twice the average loss from all diseases of swine in the state of Nebraska for the year 1891, which is given as 4 per cent by the statistical division of his department. If we correct this statement and make it accord with the letters received by the department from the owners of the inoculated herds, which letters are given in this bulletin, we must add the herd of John Campbell, which evidently was not infected before inoculation, but which plainly contracted the disease from the operation. We should also add to the losses the 2 belonging to W. Rotton, which died from exposure to cholera, the 10 belonging to Henry Dan, which probably contracted the disease from the inoculation, and the 3 belonging to W. E. Dietericks. This would give a total of 2,643 healthy hogs inoculated and a loss of 273, or more than 10 per cent. This loss is two and a half times the average loss of the state for the year from all diseases.—(Bulletin No. 8.)

Nothing can better demonstrate the fallacy of the position of Mr. Rusk and his advisers than the statement made above. Mr. R. M. Allen, of the Standard Cattle Company, Ames, Neb., writes me, "that statement alone is sufficient to show the unreliability of the bulletin." The absolute unreliability of his evidence, as given by his correspondents in various parts of Nebraska, has been conclusively shown. Were Mr. Rusk's statistics worth the paper they are printed on, did he desire to give the honest facts in the case, his statistics must have first shown that at least for six years there has not only never been so few hogs in the states but also so little cholera as in the fall and winter of 1891–92. Again, had he had reliable information from every county in Nebraska, he would have discovered that, almost without exception, inoculation was only performed in those counties and districts where the disease was prevailing to an alarming extent at the very time the owners sent for virus. *The only honest way, then, would have been to compare the loss in the inoculated herds with those in uninoculated herds in their immediate vicinity.* It was not honest to ignore my own statement, and infer if not assert, for it makes no difference with the average farmer, who would only see the loss, that the losses in herds I reported diseased at the time, or which come down within twenty days after inoculation, were due to that, when, as has been shown, large numbers of hogs inoculated at the same time with the same virus were absolutely uninjured. It was due to the farmers, if not to this station, to have made inquiries here, and published the entire facts.

Take, for instance, the inoculations done at Axtell, Neb., by Mr. Ernest Schaff, who inoculated for

D. E. Pamblade, 73 head; none sick until after January 1; loss, 30.

Gustav Lundren, 77 head; none sick; no loss.

Gustav Johnson, 89 head; inoculated September 28; sick October 25; loss, 40.

Taking those who lost, neither lost 50 per cent, while Mr. Schaff reports to me, "Adjoining farms, all around the inoculated ones, lost 95 per cent of their hogs. No hogs were hurt by inoculation." That puts an entirely different shade on the story, which is increased when we know the whole facts. My records show that the Pamblade and Lundren hogs were first inoculated September 21, 1891, merely to satisfy the demand with that virus obtained from Diller, Neb., August 8, which I knew would not protect and which, as can be seen, for some delay, probably by the express, was not used, at least as soon as it should have been. This is the virus which the reliable informants of the department claim killed the state farm hogs inoculated August 12, 1891. Now again, this case is doubly interesting, for the second inoculation of these hogs was done with a virus in the second generation (Mr. Pamblade's) and with one (Mr. Lundren's) made from the farm inoculated and at the same time naturally infected hogs, and by this, and another experiment made to control this second one, we found out that a virus from such a hog, or a hog previously inoculated and not protected, *is next to useless as a preventive virus.* We had no idea of this fact at the time. Mr. Lundren's hogs were doubtless not exposed for some reason. For the same reason Mr. Johnson's hogs did not stand as they otherwise surely would, for *they only received one* inoculation of the second generation from the state farm hogs, and now we know that a second generation is also unreliable. These statements are not made as excuses for failure, but to show their causes, that others may avoid them. We purposely used but one inoculation, in general, last year; hereafter we shall invariably use two.

It must be borne in mind that while my experiences before my return to Nebraska in 1891 had led me to suspect the various causes of failures here stated, that it was not until then that I had opportunity to exactly demonstrate them by experiment, and that the only field for experiment open to me was public inoculations on the farms, so that on beginning inoculations in August, 1891, I determined to systematically prove or disprove these points, no matter at what danger of a temporary misjudgment on the part of the public. Furthermore, I could not come to any conclusions as to results until all my returns were in, which was not until about February 1, 1892, my letters of inquiry being sent out January 12, 1892. I did not intend making any report until I had had time to fully tabulate all my experiences since 1886, with every kind of virus; history of the outbreaks from which the sick hogs yielding the virus was taken; with autopsy notes of each hog used, showing the time it had been sick, with the effect of the time the germs were in the hog on the preventive power of the virus, and lastly, the effects of time in artificial (bouillon) cultures on losing the preventive power in such cultures. Any one at all acquainted with such work can readily see the vast amount of labor it would require; but with other demands on me constantly coming up it has thus far been impossible to begin it, particularly as the regents have demanded a revision of my bulletins, with the results of recent researches, to be made ready as soon as possible. On the other hand, I cannot see what interest such tables would have to any one but certain investigators, the above statement of the results being all the public desires and really all that is necessary for the guidance of other investigators. Of course there may have been other unknown factors at work as the cause of the failures that have been hypothecated, but in this publication I have endeavored to give the conditions under which failures of any kind have occurred as well as the only conditions under which absolute success has been attained.

The accusation that the hogs inoculated by Mr. Cadwell at Ottawa, Ill., were not only killed by the inoculation, and were the cause of extending the disease to the other hogs in that experiment, and the government's warning of the terrible dangers of our method, caused me to publish the table of inoculations issued last winter, but only such parts as do show to any unbiased observer that not a hog was injured by inoculation. In justice to myself I must say that this cry of "danger" has caused such a prejudice against me that I have not only found it very difficult to get sick pigs sent to me to obtain virus from, but still more the kind I wanted, because the disease is not always so obliging as to appear in the localities where my most intelligent supporters and inoculators live, and I have had to use hogs for virus at times to meet the public demand which had been ill longer than I desired, and to thus take risks of failures otherwise unnecessary, though at the same time new confirmation of my conclusions as to the causes of partial or complete failure have been accumulated. Let me say again, that I have not yet been able to establish the safe maximal dose to prevent and not injure for little pigs under a month old, but we are approaching it.

In connection with the loss of the preventive power in other generations than the first, and proportionally, I have one very instructive failure made with the fourth generation of the most reliable virus sent out last fall in the first generation. This case did not come to Mr. Rusk's knowledge, the owner writing that "he did not answer Mr. Rusk's questions, knowing the use that would be made of them." This was the only total failure made in the inoculations of 1891, and was in the hogs of Mr. W. S. Brown, Fairmont, Neb. The fourth generation was sent to Mr. Brown because we had no other at the time, and it was absolutely impossible to get a hog sent to us that had not been sick too long to use, though we condemned six before the natural disease broke out in Mr. Brown's hogs. Mr. Brown inoculated 109 head September 18, 1891, and writes, "I return the instruments. Cholera is on all sides of me and the probability is that it will be a failure. I have a number of hogs and shoats already sick with what I call the measles; some have been that way for two weeks." As the first generation of the culture proved to be absolutely reliable, even in cases where farmers purposely exposed their hogs, this failure of the fourth generation is of inestimable value as confirming the now known fact that the virus completely loses its protective power in a fourth generation of weekly transfers from the original.

My reports not having come in, and having therefore not had the opportunity of making these invaluable comparisons, and it having been hinted by the Washington authorities that I knew the weather was getting unsuitable for inoculation on account of its coldness, and hence did my best to delay the work at Ottawa, all of which is untrue, I determined to try the effects of cold weather on inoculated hogs, and thereby unintentionally obtained two valuable results in one. I have said that virus cannot be relied on when taken from an inoculated hog that succumbs to the disease later on, and that I did not know this until I had had time to compare the results of my work. I have referred to the virus made from the inoculated farm hogs with the Diller virus, which I knew would not protect, and that succumbed to natural infection. That case was not clear enough, because the state farm pigs come down with the natural disease about two weeks after inoculation, so I determined to try it over again with an inoculated animal in which all the effects of the inoculation has unquestionably passed away and with a virus that had unquestionably given protection to the older hogs of the same herd. It has also been mentioned that it is still a matter of delicate experiment to fix the maximal amount of a single dose for quite young pigs that will prevent and still not injure or cause serious disease.

Desiring to see further and more decidedly whether my suspicion was well founded that virus from an inoculated hog that had entirely gotten over the inoculation, and in which there was no possibility of the inoculated germs being present, and also to try the effects of cold weather, and having demands for virus, I sent for a pig inoculated in September, 1891, in which disease appeared in January, 1892, and the following herds were inoculated January 8, 1892:

J. E. Brown, Wyoming, Neb., (65 four months old, 23 one year old)	88
A. R. Keim, Falls City, Neb., (five months old)	29
D. A. Baker, Ashland, Neb., (average nine months old)	110
J. E. Roe, Jansen, Neb., (two months old)	137
S. L. Furlong, Rock Bluff, Neb., (five months old)	40
John Morgan, Lincoln, Neb., (two months old)	17
Spencer Day, North Bend, Neb., (four months old)	17
Fred Schraeder, Lawrence, Neb., (seven months old)	105
J. I. Boyer, Denver, Col., (two months old)	50
	526

For ten days immediately following the inoculation the thermometer ranged from 20° to 30° below zero. Now, here is a case for the government to claim that the inoculation did some damage, for it is exactly similar to the celebrated Surprise case.

Mr. Brown, of Wyoming, writes, "the pigs were mostly in fine condition when inoculated, and were in good quarters, yet they would pile up and the mortality was 50 per cent in the small pigs and 25 per cent in the large hogs. I would say that, in my opinion, the deaths were occasioned entirely by suffocation caused by the piling up."

Mr. J. E. Roe lost six of his 137 two months old pigs from piling up, otherwise not one of the 526 ever showed any injury.

Now the question is, was it the "piling up" or the inoculation which caused the disease in Mr. Brown's, for they came down sick immediately after being inoculated. As none of the others were affected at all under exactly similar conditions, and as the disease was very prevalent in Mr. Brown's vicinity at the time, it seems to me nothing but rational judgment to assume that they were in the process of natural infection when inoculated, though the "piling up" in the extreme cold weather may have increased the mortality.

A loss of six two-months-old pigs in such a large herd as Mr. Roe's under such extremely cold conditions is a very small loss and shows unusually good care. How can we explain the absolute exemption of the others from any symptoms of illness under the same conditions? I think that settles the fact that hogs can be inoculated safely in extreme cold weather, though I don't advise it, as freezing seems to entirely nullify the power of the germs to produce this protective principle, though whatever virulence there may be in a culture is unaffected. The cold weather problem was quickly settled so far as producing an ill effect itself in inoculated hogs, but the value of the virus from a hog that had succumbed to natural disease months after inoculation could not be settled so quickly, and only the other day Mr. Roe wrote me that what I expected might be the case was in reality a fact; his pigs did not stand at all. So here is the necessary proof. *Never use a hog that has been inoculated and later on succumbs to natural infection to obtain virus from.* The cause of this must be sought in the hog and some effect on the natural germs; that is, exerted on them by the previous inoculation in hogs that directly acquire the disease that have not been inoculated.

HOGS INOCULATED IN 1891.

NAME.	No.	Injured by inoculation	Died from cholera after thirty days after inoculation.	REMARKS.
Jacob Dick	5	None		Good thing. } Full report not received for latest information.
John Morgan	17	None		Believe in it. }
J. P. Cully	44	Sick when inoculated		Lost only those sick at the time. Think it a preventive when done right. Shall inoculate again in spring, as will also my neighbors. Inoculated some for my brother. Did not hurt them. They are all right.
A. M. Caldwell	50	None	Two	Don't think it dangerous and would not think of farming without it.
Joe Antes	50	None		Disease all around me, only four rods off. Sick neighbors' hogs got in with inoculated. Don't think it dangerous. All mine did nicely.
F. H. Connelly	163	Never off feed	Lost two little pigs	Neighbor has it now. At present think it valuable.
D. E. Pamblade	73	None	Thirty	Cholera all around at time of inoculation. Loss on uninoculated farms has been 95 per cent.
Gustav Lundren	77	None		Cholera all around. Loss of neighbors 95 per cent.
Gustav Johnson	89	None	Forty	Cholera all around. Commenced to die three weeks after inoculation, and went off slowly in comparison to adjoining farms where the loss has been 95 per cent.
Spencer Day	70	None		Lost twenty-five head some time before inoculation, then put inoculated ones where these died, and they are all right. If used according to directions not dangerous and of great value, and then a perfect protection. Loss has been 50 per cent for five years.
W. H. Diller	196	None		Cholera in vicinity. Have lost heavily past years. Think it very valuable.
Dr. H. G. Leisering	50	None		Cholera within one-fourth of a mile. Consider it scientific and correct if used as directed in all respects.
F. I. Foss	153	None		Away from home, but hogs known to be all right, and that he is strong for inoculation.
Matthew & Jenner	31	None	Two little pigs died	Lost all in 1889–90. Satisfied so far, but cannot give a conclusive opinion yet.
Isaac Pollard	279	None	3 to 6 mos. old, 15; younger, 23	Older hogs all right. This shows the dose was not heavy enough in the younger ones. The others resisting and still among them speaks for inoculation as strongly as could be desired.
Mr. Abbott	79	None		Disease on place at time. Can't find this man by mail. Is an invalid and may be away.
P. J. Donlan	100			All right.
A. B. Wright	75			Disease on place at time or soon after. I inoculated sows and pigs at same time—pigs died—*all the old hogs lived.*
J. D. Steiner	57	None		Lost fifty head in '90, and wasted $50 on medicine. Disease in vicinity. Think it very valuable.
J. B. Diller		None		1885 to 1890 lost $2,000 worth. Valuable and safe.
John Campbell	85			Sick at time of inoculation and lost sixty shoats, but none of the old ones. Worthy further trial.
W. S. Brown	108		Seventy-seven	Fourth generation used, as could not get right virus and disease very near. Inoculated sows and sucking pigs at same time by mistake.
Ed. Slattery	7	None		Lost previous years. Disease in vicinity. Think it a success.

Chas. Argyle	24			By mistake inoculated sows and sucking pigs at same time. Seven of the pigs died; think they got a double dose. Loss on this place has been so great that it was given up and I hired it. Disease all around. Don't think it dangerous. A man should inoculate every year, then he won't make a mistake. Sows and pigs must not be inoculated at the same time. Think it all right and just what we want.
Fred. Braasch	106			Disease quarter mile off. Inoculation not dangerous.
Ezra Wilter	195	None		Loss fifty per cent two previous years. Disease all around and sick hogs got in with mine. Same faith as in vaccination in my family.
C. H. Crocker	9	None		Cholera been on place for years and now in vicinity. Shall continue to inoculate all my hogs when on infected land.
Charles Cline	28	None		Quite a loss last three years. Neighbors' sick hogs got in with my inoculated ones. Valuable and not dangerous without limitations.
Samuel Lewis	118			All right.
S. M. Geyer	60	None		Cholera near me. According to general report seems to be dangerous, but has done mine no harm.
C. B. Boylston	68			Lost all but inoculated hogs.
Hugh McLaughlin	40			Hogs sick at time of inoculation; lost fifteen head. Lost fifty head annually since 1889. Think inoculation saved me a greater loss.
Michael Quinn	62	None		Lost badly previous years. Lost three hogs from cholera that were not inoculated. Think it safe and a good thing.
Peter Yunker	40	None		Not sick a day and doing splendidly. For several years lost practically all I had.
A. B. Seymour	48	None		Cholera in vicinity. Best thing yet discovered.
Henry Dan	49	None		See previous consideration.
W. C. Dietericks	44	None		See previous consideration.
W. Rotton, Jr.	53	None		See previous consideration.
H. H. Hyde	34			Lost heavily previous years. Disease all around—only twenty rods away. Not dangerous, and a good thing.
Geo. Young & Son	30	None		Disease within a half a mile. Worth a fair trial. Not experience enough yet to give opinion as to value.
J. R. Alexander	40			Sick at time of inoculation. My loss has been less than any of my neighbors. Think it our only hope.
John Hart	13			No fair test. My hogs were sick when inocuulated, so can't say as to value.
S. C. Bassett	16			Inoculation do s not kill, stunt, or injure it done as directed. Could see no ill effects whatever. 1887 lost 75 per cent. Spring of '91 lost 60 out of 80 head of uninoculated, while other hogs on place that had been inoculated were not affected all. Feel sure that had those which died been inoculated that they would not have contracted the disease. Have practiced inoculation since 1888 with this one exception.
Joseph Gross	18			Sick at time of inoculation; visibly so; 13 not sick; these 13 never got sick. Altogether my neighbors lost as follows: one man, 32 out of 37; another 28 out of 35; another 93 out of 94

The government asserts that inoculation is a dangerous procedure, that it not only kills and stunts the hog, but extends the disease. Among other things our last year's work was especially directed to show not only that it would not injure hogs, but that the farmers could be trusted to inoculate their own hogs, and I am now trying, and slowly succeeding in getting them to make their own virus if we send them the bouillon. The foregoing table is a truthful statement as to injury done the hogs and shows absolutely that not a case of loss can be ascribed to inoculation. The numerous letters published show that no one who has had experience thinks that inoculation ever injured a hog, unless too closely confined.

The losses of Messrs. Dan, Dietericks, and Rotton have been fully considered in another part of this paper, where it has been shown that by no possible twisting of the evidence could they be attributed to inoculation. The work of the Agricultural Department has, however, led men to attribute losses to inoculation in herds that were never inoculated and in counties where no inoculation has ever been done.

Whether or not the inoculation done by Mr. Cadwell at Ottawa caused the disease in his hogs, or whether they caught the disease from the government hogs, I can't positively say, though I believe neither to have been so. To me it seems not only impossible, but that opinion is supported, as has been shown, by the advice and experience of Mr. Rusk's advisers. Nor can I say how those hogs caused the disease in the others any more than the government in theirs, nor why it should have in either, when it did, if they were properly isolated. For four years now we have constantly had inoculated and uninoculated hogs in the same building, the same man having charge of them, and not once has disease appeared in our uninoculated stock. I purposely inoculated six one-month-old pigs with one-half cubic centimeter (double the dose generally given) of a culture but three days old, from such an outbreak as I desired to make virus from; they are penned and not in an open lot with free range; about the tenth day several of them had a little diarrhea and were off their feed, and one died the nineteenth day with scarcely any lesions present; and another August 12, having both pulmonary and intestinal lesions; the others came round all right in a few days and are doing as well as ever again. *Four* uninoculated pigs in the same pen with them and subject to all possible exposure have never lost a meal, which only confirms a long list of similar attempts to confer the disease to uninoculated controls by the hogs inoculated with a properly selected virus. Where, then, is the danger? The danger lies in not following the instructions given in every detail, for which we cannot be held responsible. As to extending the disease, we use every precaution posssible, as can be seen by the following directions:

PATHO-BIOLOGICAL LABORATORY, STATE UNIVERSITY,
CORNER 10TH AND K STREETS, LINCOLN, NEB.

Directions for Inoculation.

1st. *No hogs or pigs should be inoculated except on premises on which swine plague has prevailed in previous years.*

2d. *It is absolutely useless to inoculate swine when already diseased, as inoculation of hogs, like vaccination of children against small-pox, must be done before the disease attacks them.*

3d. While the virus and necessary implements will be supplied to any farmer or breeder in Nebraska free of charge, except that such farmer or breeder must pay the express charges on the same from Lincoln, and for the return of the implements to Lincoln, *still any person who neglects to return said implements, will be shut off, and in future shut off, from the privilege of obtaining virus from this laboratory.*

4th. In the box sent will be found a flask containing the virus, which must be used within the limit of time designated. Shake the flask well before using. Pour out some of the contents into a clean tea-cup or some convenient vessel.

5th. With each flask will be found a glass syringe with a cap on the end. Unscrew the cap and fill the syringe with hot water and let it stand a few moments, or fill it several times; then screw on one of the accompanying needles and squirt out the water; then prove the syringe by filling it through the needle, and after squirting the water out, fill it with virus from the cup, and it is ready for use. After using, wash out with hot water, unscrew needle and put the wire in needle again, and screw the cap on the syringe.

6th. THE DOSE.—For pigs three to four weeks old, one-quarter of a syringe; three to five months old, one-half syringe; old hogs, one syringe, unless otherwise specially directed. Second inoculation twenty days later, and double the quantity in each case.

7th. HOW TO HANDLE THE HOGS.—Young stock can best be handled by lifting them by the hind legs and holding them between the knees. Old stock should be laid on the side and strongly held. Care must be taken not to use violence, and thus make lame the pigs. This done, introduce the needle of the syringe through the thin skin of the inside of the hind leg, and then push it until in, along under the skin, and *not down into the flesh;* then squirt in the indicated amount of virus, and let the hog go. One hundred can be easily done in an hour, with the necessary help to catch and hold them.

8th. No change of food is necessary. Do not pen up inoculated hogs.

9th. The stock cannot be considered to have been safely inoculated until twenty days after the second inoculation.

10th. Return the box and implements as soon as possible after using.

11th. Be sure and fill out the enclosed register.

FRANK S. BILLINGS, M. D., *Director.*

Restricting our work, as we peremptorily do, to the inoculation of hogs only on farms where losses have occurred in previous years, and where the disease is expected to appear at any moment, to which must be added the almost universal fact, that farmers, even on such farms, do not apply for virus to inoculate their hogs until the disease is very near them and often not until they are suspicious of the condition of their own herds, it is evident that, no matter who the party may be supplying the virus, cases will occur in which infected herds will be inoculated, and that the more inoculation is done the more numerous will such cases be found. It is such cases as this that Mr. Rusk asserts I have reported falsely about. He assumes, because it suits him, that all such cases were caused by inoculation. He positively refuses to consider the fact that we may have even ten or more herds of 1,000 or more hogs all inoculated with the same virus at the same time, and every hog may not only be well but never ill since inoculation, and yet he does assert that inoculation killed the hogs in the one case. It matters not to me if half the herds in such a case come down with the cholera within twenty days after inoculation, if the other half, or even but one good-sized herd, inoculated as they were at the same time, shows no sign of illness, I dare affirm, and know every honest investigator will support me, that inoculation could not possibly have injured the others. I simply say—*it is impossible.*

I am not one to "cry baby," but I do demand cold-blooded justice. With Mr. Rusk and his aids I neither want nor shall I give quarter. But I will be just and demand justice. This thing does not trouble me except that it has stampeded the farmers so badly that, as can be largely seen in the letters published by Mr. Rusk, men have asserted that hogs which were never inoculated were killed thereby, and as no correction of that statement is made in Bulletin No. 8, it is evident that the secretary of agriculture absolutely has no desire or intention of treating this matter justly. Again, if we inoculate eight or ten bunches of hogs and one comes down a few days later with disease, even though the owner writes us he is "sure they must have been sick at the time, for all the other herds done the same day in my vicinity are all well and don't seem to have lost a meal," yet he goes to town and sees some men who have read Mr. Rusk's dissertations, and the next thing we hear that we are the "biggest fraud on earth and inoculation killed all my hogs."

All I have to say to Mr. Rusk is, "to let the farmers alone and treat them one-half as honestly as we do, and see if he cannot do better and not try to undo for us our honest attempts to get at the bottom of this problem.

Any one with common sense knows that inoculation cannot make an already infected farm any more dangerous to the hogs on it, or to those in the neighborhood, than it already is. Safety consists in the neighbors keeping out of such an infected hog lot, and in the owner keeping dogs and such things out of it. The question to that farmer is, *how can I raise hogs on my farm?* That is all I care about. Take, for example, the following letter from one of the largest cattle and hog feeders in Nebraska:

SHELBY, NEB., Jan. 12, 1892.

MY DEAR DOCTOR: I inoculated fifty hogs a year ago last fall and only lost two that winter. I think inoculation the only thing that will save them, and I shall inoculate in the future. I did not inoculate this fall and I lost nearly every hog I had. I have lost hogs from cholera every year for the past eight years, except the one year I inoculated, all in the same feed yard, and I have confidence enough in it to inoculate from now on, as it seems to be the only reliable preventive.

Yours truly, TEMPLE REID.

There are thousands of just such cases in these western states. Now, if Mr. Reid's experience in that one year can be repeated, and we have several hundred exactly similar cases on record, is there not every reason that such a farmer should take advantage of inoculation? Is there any sense or justice in Mr. Rusk's wolf-cry, "danger," if inoculation is only done on such farms? Mr. Rusk himself gives exactly the same advice in his report of 1890, and does not even advise the precaution we insist on, of limiting inoculation only to infected farms.

He even recommends our method (except in the selection of the hog to take virus from), though he claims and demonstrates his investigators have never successfully inoculated with any method they have tried. He says:

The method of subcutaneous injections of culture liquids containing hog cholera bacilli, while on the one hand fraught with the possible danger of scattering diseased germs where they do not originally exist, is nevertheless the simplest and cheapest method that can be devised for the vaccination of animals; these qualities of simplicity and cheapness are of vital importance in a question which has only a commercial aspect. (P. 111.)

Now, I beg to say that to recommend the very method, so far as use is concerned, practiced here, regardless even of the one precaution which we consider necessary to keep the disease restricted to farms already infected, and then to decry that same method as dangerous merely because it succeeds with us when done rightly, and fails with his men because it never had been done as we do it and succeed, is the height of inconsistency.

I wish to call the reader's attention to a most striking inconsistency between a passage in the above quotation and Mr. Rusk's advice, that swine plague is a contagious disease and can only be stamped out in the same manner as is applied to contagious pleuro-pneumonia, in Mr. Rusk's previously quoted letter to the *National Stockman*. In this report of 1890, in the passage above quoted, he says:

That while the subcutaneous inoculation of culture liquids of the hog cholera bacilli is on the one hand fraught with the possible danger of scattering disease germs *where they do not originally exist?*

Where is that point? It must be the land. But by his own letter of advice to the *Stockman* he insists that they only originally exist in the hog. Now if they only originally exist in the hog there is not a contagious disease germ on earth that is known to live for any length of time when exposed to the elements by being scattered unprotected on the surface of the soil. This knocks the "contagious" disease argument at once in the head.

One more point and I am done with this part of this paper: Mr. Rusk and his advisers assert that this work should not be done by the farmers and cannot be safely entrusted to them. We say here, that if it cannot be done that way it cannot be made practical. I say it can, and think it has been demonstrated that I am correct. I am bitterly opposed to advocating anything unnecessary that will increase the number of office holders and political ringsters at the disposal of party leaders. I know that the most intelligent farmers in our communities, and they are at present the very ones taking hold of inoculation, are fully as capable of being trusted with every detail of this work as a small army of political veterinarians who, so far as swine plague goes, know nothing about it in comparison to our intelligent swine breeders. I can see only public danger in increasing the number of veterinary parasites at the command of the politicians in the Agricultural Department. *The great question is to educate the people to do all they can for themselves, and to carry on the government with the greatest possible economy by cutting down every unnecessary man in the public service.* It is to the interest of the politician to do just the contrary, for in an army of such parasites lies his political power as a boss-heeler.

To the political boss, this country is a government of the people, by political ring masters, for their own benefit, instead of a government of the people, for the people, by the greatest intelligence among the people, selected and elected by the people.

Knowing the people can be educated to inoculate their own hogs, knowing they can do it successfully in time, and knowing that to put it in the hands of an army of veterinary parasites would only unnecessarily increase the tax burdens of the farmers, I have drawn up the following instructions for them to study:

EVERY INTELLIGENT FARMER CAN PREPARE HIS OWN VIRUS AND INOCULATE HIS OWN HOGS.

That is what I say. The government says quite the other thing. In its latest attack on our endeavors to aid the swine dealers of the country, but especially those of Nebraska, the government says it does "not believe that inoculation could be safely trusted to the use of the average farmer, or, for that matter, to the average veterinarian."

This statement is like the majority of those emanating from the same source, absolutely without foundation and valueless, and is in fact contradicted by the Chief of the Bureau of Animal Industry in his letter to the editor of the Iowa *Homestead*, where he says:

And in the report of the bureau for 1886, pages 60 to 70, the details of the experiments are given so fully that any one who can make a culture of germs can repeat the experiment for himself.

It is contradicted by the actual experience of hundreds of farmers who have inoculated their own hogs and are still doing it, and have never seen a single case of injury to their hogs. That statement is so absurd that it needs no further consideration. I now intend to go further than I ever have, as I am confident in the correctness of my statements, and desire that the farmers of Nebraska shall learn to make their own virus. In other words, I propose to supply them, as fast as they can be instructed, with the uninoculated beef soup bouillon ready for use and the platinum wire to inoculate the soup with. Having these things and the syringe and needles on hand and following the directions to be given (instruction will be given at the laboratory to those who desire it) inoculation can be made a practical success, and need never fail. The best proof of the value of a thing is its simplicity.

We will assume that Mr. Charles Walker (who knows all about it), one of the best known farmers in Nebraska, has the vials of soup and the wire in a glass rod on hand and desires to inoculate his own hogs. What does Mr. Walker do?

First—He looks around for outbreaks of swine plague and selects the mildest one he can find, and above all things avoids one that is killing a large number of hogs in the herd and doing it anywhere from one to ten days. The greater the number of deaths in a herd, and the shorter the period of illness, the more unsuitable is an outbreak to obtain virus for inoculation from. Whereas the smaller the number of animals ill, the slower the course of the disease in such, the better is such an outbreak suited to obtain virus from.

Second—From the last kind of an outbreak Mr. Walker selects a pig or hog just taken ill, and not one that has been sick some time (remember this: the animal to be taken must not have been sick long, for the sooner after it is observed to be ill the virus is taken the more reliable will it be; chronic cases are useless; no dead ones must be used), and kills it by a rap on the head—not by bleeding. We must keep all the blood in the animal. He lets the animal lie until cold, and then takes a good knife and cuts open the skin from between the jaws along the belly to the tail. He then cuts the skin away from the body down both sides, but in the vicinity of the forelegs cuts away the latter carefully, for if he cuts too deep down, where they are attached to the body, he will cut the large blood vessels which enter the forelegs, and thus bleed the animal, which he does not want to do. He then cuts the abdomen open in a straight line from the posterior end of the breast bone to the hind legs, but is careful not to cut the intestines. With the abdomen or its contents he has nothing further to do, but he must open it in order to open the breast cavity, which he does by cutting across all the ribs and muscles about two inches each side of the breast bone, which he then separates from the diaphragm, a partition between the chest and abdomen, and lifts off the breast bone and bends it back towards the head, thus exposing the contents of the chest, but especially the heart.

Third—The heart has two main chambers, called ventricles. The one, the left, has very thick, solid walls; the other, the right, has thin and flabby walls. He twists the heart a little so as to bring this thin side up. He then puts some absorbent cotton or a piece of sponge into a tin box and fills it full of alcohol, which he now sets on fire. Naturally, he must have his hog where the wind does not blow, and as no blood need be let out, it can be done on the kitchen table. After lighting his alcohol, Mr. Walker takes a cheap kitchen knife and heats it hot in the flame and then burns over the outside surface of the right side of the heart, the thin wall. Next he takes a small knife, and this must have a thin blade, and heats that and then cuts and burns a slit through the thin wall of the heart until the blood flows out.

Fourth—How to inoculate the beef soup: It has been said that the laboratory will supply farmers with this soup in flasks all ready for use, and also with an inoculating wire and syringe. The inoculating rod is made of a piece of platinum wire fastened into a glass rod. Mr. Walker now passes this glass rod a few times through the alcohol flame and heats the wire, in its end, red hot, then he lets it cool a moment and removing the cap from the flask, he loosens the glass stopple; then he dips the point of the wire, which has a little loop in its end, into the blood in the heart through the slit he has cut with a hot knife, and quickly introduces the wire into the soup, putting the stopple in at once. This he repeats three times, removing and replacing the stopple each time. He now puts the glass cap on the flask and sets it in the kitchen in a safe place for three or four days. Each flask contains soup enough for 100 grown hogs. The germs at once begin to multiply, and soon the previously clear soup becomes clouded and milky, if the germs are in it, which will occur in nearly every case. At the end of three or four days it is ready for use and should be used as soon as possible after that.

Fifth—Remember this: The first generation is the only reliable virus. No other or older generation should be used. They are not reliable. Again, the first generation should be used between the fourth and seventh day after inoculating the soup from the heart's blood of the sick hog, *never earlier and never later.*

As every sick hog but one that was used to make virus from was selected and sent to me by farmers, it is evident that these men, at least, have appreciated the one fact, and that is, *to avoid danger and not select a hog from a malignant outbreak* as was said to be mistakenly done at Ottawa, Ill. The proof that they made no mistake in that direction is to be seen in the results. The few cases in which farmers made the mistake of inoculating sows and pigs at the same time must not be counted as an example of the dangers of inoculation, because it was thought unnecessary to print directions on that point. In each of such cases it can be shown by the unabbreviated statistics at the laboratory that several hundred hogs were inoculated in the same locality, at the same time, and with the same virus, without a particle of injury. These few mistakes on the part of farmers show that we must add the following to our list of instructions:

SPECIAL CAUTION.

On no account inoculate either piggy sows or sows suckling pigs. Don't inoculate sows and pigs at same time. Wean pigs first, unless sows have been inoculated, or had the disease and recovered, because the sows may take up the germ passed off the pigs, and then it will get into their milk, and, though not strong enough to injure the sows, may kill the pigs, as they would then get an unlimited dose. Do not confine inoculated animals.

What mistake have the farmers made that have sent in hogs? Only one. Either through economical reasons, or not careful observation enough, or wrong information from the actual owner of the sick hog, quite a large number have been sent in that have been ill too long to be useful for virus. *The shorter time a hog is sick the more reliable the virus made from it will be, and the less the danger of other germs being there;* though in most cases that would not harm the virus. I will assume all risks from the presence of the government swine plague germ doing any injury if in the virus.

How can a farmer tell what hog not to use?

Cut the hog open as above described: if the lungs are badly diseased, very solid, and attached to the ribs, throw it away; if the lungs are only slightly touched, make the virus from the heart's blood as directed; then cut open the large gut and wash it, and if there are no blackish looking ulcerations or hard tumors in it, go ahead and use the virus; if there are, condemn it and get another and fresher outbreak; that hog has been sick too long.

Any intelligent farmer can himself learn all this by cutting open a few hogs, or all that desire will be cheerfully taught at the laboratory.

The above is every precaution that has been resorted to in the laboratory; that is, only such as the common every-day farmer must and can use. No laboratory methods have been used previous to sending out the first generation of a virus. Then we have resorted to such methods and *rabbit tests* in order to see just what the farmers would have in virus made that way. In the majority of the cases the virus has been pure; in some one or two, harmless germs have been mixed with that of swine plague in small numbers; in two, the government second plague germ was present, but as in those cases the hogs are still alive and were never sick, its terrible dangers are manifested: *it will kill rabbits.* The only trouble that has been had has been to get sick hogs frequently enough, so that much against my desire the second and third and in one case the fourth generation has been used, a thing that will never be done again, especially during my absence in Chicago. So far as inoculating the bouillon from the sick hog at the laboratory is concerned, I purposely left it to my assistant, who simply followed the instructions herein given and does not know as much about swine diseases as many a farmer. The results are now before the public, and they can judge. Further comment from me is unnecessary.

PURE CULTURES NOT AN ABSOLUTE NECESSITY FOR PRACTICAL INOCULATION.

The objection has been made that farmers will not have skill enough to obtain pure cultures, and the inference made thereby that pure cultures are absolutely necessary to successful results. The authors of such assertions should know better even if they do not. All practical and experimental experiences speak to the contrary. For instance, admitting for the moment that farmers will not always obtain pure cultures; admitting for our purpose that Mr. Cadwell's inoculation did cause the disease in the hogs at Ottawa, Ill., by subcutaneous inoculation as the government claims; admitting even, what we do not know, that his cultures were pure; still, taking the government statement to be correct, *have we not conclusive evidence that a farmer can make a culture from sick hogs that will produce swine plague?*

Although for the reasons previously given I do not and cannot admit that Mr. Cadwell's inoculations was the cause of the fatal disease in those Ottawa hogs, still, if the government is correct, we have evidence that Mr. Cadwell knew what to do to obtain cultures from a diseased hog and to inoculate and prove he had the specific germ in his cultures. This is evidence enough that a farmer can be taught to do the work, and we shall more fully demonstrate this in future. On the other hand, an experience of over six years shows that if the selection of the infected animal from which virus is taken is carried out according to the simple directions so frequently given, viz., the most recently infected hog in a very recent outbreak, and the simple method of sterilization, which anybody can acquire in five minutes' instruction, carefully followed, that in nearly every case pure cultures must result, for the simple reason that those destructive lesions which give openings for the entrance of adventitious germs in the blood have not had time to be developed.

Again, the very lesson and value of rabbit or small animal inoculation, and the use so commonly made of them to obtain a pure culture by the subcutaneous inoculation of mixed cultures, in order to save time and trouble, is as well known to the opponents of farmer inoculation as to myself. The small animals die; the unwished-for germs remain local at the point of inoculation, while the specific germs find their way quickly into the blood and kill the animal, from which we obtain them in pure cultures. The same thing is as sure to take place in the hog the farmer inoculates.

To be sure the "bug-a-boo," that the germ of the government's second plague may be present, has been brought forward, in an endeavor to frighten the farmers against inoculation; but if the government's publications can be relied on, and this time they are confirmed by our own observations, that that germ is simply a common organism, not only in the air passages of swine, but other domestic animals, and can only gain entrance to the blood after severe disturbances have developed in the lungs, *then, following implicitly our directions excludes that possibility, which fact is confirmed by a very great number of rabbit and pig tests with such blood, and also by means of that most exact of all methods, plate cultures.* Again, it has been demonstrated by a great number of inoculation and some feeding tests on hogs, here published, that that government germ is by no means the specific, dangerous factor it has been represented to be in order to alarm the farmers of the country, as is shown by the following:

In a late publication on swine diseases by Professor Welch, of Johns Hopkins, he asserts that not a single undoubted epizootic outbreak of this second swine plague has yet been seen in this country.

And Salmon said, in his address before the National Swine Breeders' Association, "then there is the germ in hog cholera, a disease widely distributed, generally extremely fatal, *and probably productive of the greater part of the loss which falls upon the hog-raisers.*"

After such an admission, is it not self-evident that the government is misrepresenting when it talks about a "widespread epidemic disease" which it calls "swine plague"?

It is self-evident, from all the assertions made by the Agricultural Department in its various publications for the past five years against our work, that they are more afraid of the public becoming aware that swine plague can be prevented ;by inoculation, and thus completely expose the futility of their own attempts in the same direction, than anything else. Cowards and frauds are always more afraid of the truth than all else in the world. That an impure cultivation can be used sucessfully as a preventive virus is proven by referring to a frequently quoted passage from the report of 1883, that

M. Pasteur has recently confirmed our American investigations in a very complete manner. He shows that the desire is produced by a micrococcus, that it is non-recurrent, that the virus may be attenuated and protect from subsequent attacks, and he promises a vaccine by spring.

Now, it is a fact, which the readers of the objections of the government, that a farmer cannot be trusted to make the virus because his cultures may not be pure, should know, that

First—*M. Pasteur did not study any swine plague or disease known in this country, either in the past or present, but the rouget or erysipelas of hogs.*

Second—*That a micrococcus is not the cause of rouget.*

Third—*That M. Pasteur never knew what germ was the cause of that disease, or that the real germ was in the very virus with which he did produce immunity, though mixed up with a microccocus and some other germs, as demonstrated by Schütz in Berlin, 1885, in Pasteur's "Vaccine Contra Rouget."*

Fourth—*That Pasteur did keep his promise, which is more than the Agricultural Department has done, and did produce immunity with a virus containing mixed cultures.*

Again, no one dare claim that the most reliable of all known immunity-producing viruses, vaccina against small-pox, is or can be a pure culture of the germs of that disease when collected on points, or in any other way, from the vesicles in the skin of calves, or from the scabs taken from the arms of vaccinated people. No one knows positively what the specific germ of small-pox or vaccina really is, and yet it must be in the above named material or there would be no disease and no immunity produced. A fact worthy of attention which it has not received is this, *it is more than probable that were the virus used for vaccination not an impure culture that scarification vaccination would be absolutely null and void; for the reason that the germ of small-pox and vaccina, self-evidently an anærobic germ, does not require and even will not develop in oxygen, and that the so-called adventitious germs polluting the virus and in the scab, consume the oxygen and thus produce the necessary conditions in which the germ of vaccina only will develop and produce the desired effect.* I could produce very practical evidence in favor of this hypothesis were I permitted to publish a series of experiments with what is undoubtedly the germ of vaccina, the property of an esteemed friend and colleague, who is unfortunately too ill at present to publish his very valuable and interesting researches.

It seems to me that the above remarks entirely answer the objections that intelligent farmers cannot be taught to make their own virus and inoculate their own hogs. It seems to me an insult to the intelligence of our hog-raisers to assume that a small army of veterinary office holders, with little or no experience in hog diseases and no better capable of being instructed in the simple technique of obtaining cultures from the heart's blood of diseased hogs, is necessary that the work may be done well. From past experience I much prefer to trust the reputation of inoculation to honest and fair-minded farmers who desire success than to government veterinarians whose chief aim seems to have been to do their utmost to make it a failure.

The following "open letter" to the chancellor of the State University appeared after this manuscript was first completed, but is here inserted so as to give the public all the documentary evidence possible. Its character can be easily seen by the reader. It is a disgrace to its author, an insult to the truth, and a deception attempted on the farmer [Chancellor Canfield's letter should in courtesy have been published also by Mr. Rusk]:

OPEN LETTER OF SECRETARY RUSK TO CHANCELLOR CANFIELD.

DEPARTMENT OF AGRICULTURE,
WASHINGTON, D. C., July 11.

To Mr. James H. Canfield, Chancellor of the State University, Lincoln, Neb.

SIR: Senator Paddock has just referred to me, for my views thereon, certain statements made in your correspondence with him in regard to Farmers' Bulletin No. 8, issued by this department, which statements are coupled with the expressed desire on your part to know "whether the head of the department has really given the matter any thought at all." As you expressed your views in regard to this bulletin quite fully in your letter to me of the 3d ultimo, in which you cover all the points which have been brought to my attention in your correspondence with the senator, and also all contained in the newspaper letter of Mr. Charles H. Walker on the same subject [see p. 144], a copy of of which you enclosed for his information, the whole subject can be sufficiently considered by transmitting an answer to your letter to me through the hands of Senator Paddock.

Your letter would have been answered sooner had you not stated explicitly that you wished me to understand that it was not to be the ground of lengthy or argumentative correspondence, and that you would be satisfied to know that it was in my hands and receiving due consideration. As, in spite of this statement, you have not been satisfied to leave the matter in that condition, I shall now reply at such length as seems desirable to demonstrate the misleading and inconsistent position which you assume, and the untenable character of your suggestions.

In your letter to me you complain because Bulletin No. 8 was not submitted to you for revision before it was printed, and you criticise the statements made in some of the letters from citizens of Nebraska which were published in that document. Under any circumstances your complaint might be considered as extraordinary, since it has never been and is not now the custom for this department to submit its reports or bulletins either to the officers of the state universities or to those of the state experiment stations to be revised or modified in advance of publication. I am not aware of any reason why this should be expected. The reports of the Nebraska experiment station certainly have not been sent to me for revision or modification, although the relation of the experiment stations to this department might readily suggest such a course, and some of the bulletins issued from that station bear upon many pages the evidence of needed revision.

So far from adopting such a friendly course, the publications issued from your patho-biological laboratory are filled with the most glaring misrepresentations of the valuable scientific work done by the Bureau of Animal Industry, and apparently no effort has been spared to express disparagement in the coarsest and most offensive language. After these assaults, inspired by jealousy and egotism, had been carried on through the bulletins of your station and for the newspapers for years, the director of the patho-biological laboratory assailed me personally in similarly abusive language because I would neither turn over a part of the bureau's appropriation for his use nor permit him to dictate as to the manner in which the investigations of this department should be conducted. Considering these facts, you need not be surprised at my astonishment when you seriously propose that I submit the bulletins of the Bureau of Animal Industry to be revised at your station in advance of publication. If a more preposterous proposition was ever made to the department it has certainly not been brought to my attention. I might have applied for information in regard to the inoculations made in Nebraska as you intimate, but if the malignant hostility shown toward this department by the person in charge of your patho-biological laboratory had not prevented me from applying to that source, the unreliable character of the statements issued by him would have been sufficient. Comparing his statements made from time to time with letters received from the unbiased citizens of Nebraska, I have no hesitation in saying that even if the latter were made from memory they deserve the greater confidence.

It must have been evident to you when your letter was prepared that your criticisms of the bulletin in question did not touch any essential part of it, nor did you point out any inaccuracies or errors in the statements made by its author. The greater part of your communication is devoted to proving that there were incorrect statements in four of the letters published which were written by Nebraska farmers. As there were about fifty-five letters printed, most of which were written from memory, the writers must have been unusually accurate if there are only four of them in which you and Mr. Walker can find statements to criticise. [A dangerous admission for a government official to make. B.]

Where you both misrepresent and try to mislead your readers is in the studious endeavor to convey the impression that the letters mentioned were used in drawing the deductions of the bulletin as to the failure of inoculation. For this there can be no excuse, as the object of publishing this correspondence is very plainly stated, namely, to show that there had been many herds inoculated during 1888 and 1889, and that many losses from inoculation occurred during those years, of which the public up to this time has had no information, and to indicate that the sentiment among the farmers in the districts where inoculation has been most thoroughly tested is overwhelmingly against the practice. The first two conclusions cannot be contested, and from the correspondence of the department I feel sure that the last one is equally correct. It may suit the purposes of the half dozen persons who are interested in sustaining inoculation to draw the attention of farmers from the facts which prove it to be a humbug by assuming such unbounded indignation over errors in a few letters which could be excluded without affecting in the least the general conclusions of the bulletin. But unless I am greatly mistaken as to the intelligence of Nebraska farmers, this plan of throwing dust in their eyes will not succeed.

What I fail to understand is, how you, an educated man, accustomed to the examination and analysis of written and printed documents, could deliberately ignore the evidence of the failure of inoculation as presented in the bulletin, and, selecting this correspondence for your text, ask me: "Is this the best evidence that can be secured for the establishment or the overthrow of scientific experiments? Is this a scientific method of investigation?" With the bulletin before you, did you not know that the necessary evidence from scientific experiments was contained in it? And if you did know it, what was the object of asking such misleading questions.

Concerning your intimation that letters favorable to inoculation were omitted from the bulletin, I have only to refer you to the letters of such men as J. W. Coulter, D. P. Ashburn, S. C. Bassett, Hugh Gibson, and Thomas Peifer, all of whom state that they are believers in inoculation, and whose letters appear in the bulletin [see Suppressed Letters, p. 121]. There were a large number of letters, both for and against the practice, which were omitted because they were written entirely from a theoretical point of view, and made no reference to any facts in support of the position taken, or for other equally good reasons.

One of these, signed by S. W. Perin, the foreman of your state farm, was so evidently written with the intention to deceive the reader that it was not given, and out of consideration for your state it was not exposed. [See page 198.] Mr. Perin does not hesitate to make the positive assertion that no hogs had been lost on the farm after inoculation, and yet I feel sure you will not question the statement that in August of last year forty-eight head of swine were inoculated on the state farm, thirty-eight of which afterwards died from an outbreak of disease set up by the inoculation, and that three out of four not inoculated contracted the disease from the inoculated animals and also died. Can you explain why this untruthful statement was made by the foreman of your state farm?

You are quite right in your assumption that I am not willing to let untruths or half truths appear on the record of the department during my administration. Are you equally particular in regard to the statements made from your patho-biological laboratory? If so, how can you sustain the director in his assertions that inoculation has been an unqualified success? Why do you ignore the main points in Bulletin No. 8, which proves inoculation to be a failure and dangerous to the stock interests of the country, and confine your criticisms to details of correspondence which do not affect the general conclusions a particle?

You admit that the inoculation of the Hess herd was a failure; but when you assert that "No one has ever denied that it was a failure," you certainly are in error. Did not the director of your laboratory assert before the National Swine Breeders' Association, November 14, 1888, that his assistants had inoculated 1,000 hogs in Nebraska, and that there had been no failure? Did not the same person assert in his first pamphlet on inoculation, published about a year later, that over 1,000 hogs had been inoculated in Nebraska since 1886, with a reported loss of but eleven out of the whole number? Did he not say, over his own signature in the Omaha *Bee* under date of January 7, 1892, that "Every one who is acquainted with the true facts knows that those herds reported as killed at Surprise, Neb., in 1888, were all diseased at the time they were inoculated"? How do you harmonize these different statements with each other, or with the letter of Mr. Hess which states that the hogs were perfectly healthy when inoculated, and only showed sickness eight or ten days afterward, or with the explanation which you now desire me to make, to the effect that the inoculation was a failure, but that "it was an early experiment, and was to be weighed as such"?

It is these inconsistent and, in some cases, plainly untruthful statements, which have emanated from the patho-biological laboratory, which have caused me to lose all confidence in the work which is done there, or the records which are kept of it. For this reason, if no other, I should decline to have the bulletins of this department modified so as to agree with those records.

But these are by no means all the inconsistencies which I might point out. I will only take the time to refer to one other. In the letter in the Omaha *Bee*, already mentioned, it is stated: "This year over 3,000 have been inoculated in Nebraska. * * * Of the 3,000 I do not know of one being injured by inoculation, though one such case in sucking pigs is reported, and one failure in the same herd," etc. At the time this was written the writer certainly knew of the failure on the state farm, and within five weeks he published a statement from his records admitting a loss of 198 head as having occurred in herds which were healthy when inoculated. Does it not occur to you that it might be well for you to experiment in revising the statements made from your patho-biological laboratory before you undertake to edit the bulletins issued by this department?

Again, you are indignant because there is a "blank silence" in the table on page 11 in regard to the experiments at Gibbon, although any one can see from a summing up which follows the table that no losses were counted against the herds located at that place. If you were examining this subject impartially, why did you not call my attention to the fact that the author of Bulletin No. 8 omitted to give the numbers that were lost in Mr. Hinckley's herd? How does it happen that in all of the efforts that have been made to elucidate the question of inoculation and to enlighten this department you have failed to quote your records in regard to this case? From a bill introduced in the Nebraska legislature asking compensation, I learn that Mr. Hinckley's loss was eighty head. This would make the total loss 543, instead of 463 as given on page 11 of the bulletin—the percentage of loss being 53½ instead of 45½. In other words, the failure was even more disastrous than was claimed in the bulletin.

Another example of generosity on the part of the author of Bulletin No. 8 is seen in his summing up of the losses which followed the inoculations made under the direction of your patho-biological laboratory during 1891. In this summing up (page 37) the herds which it was claimed were affected before inoculation were excluded from the calculation, but in a comparison of the loss among inoculated and uninoculated herds in the state, it is plain that no such exclusion should have been made. Taking all the herds inoculated in 1891 from which figures have so far been given, and I find the loss foots up 12½ per cent, instead of 10 per cent, as given in Bulletin No. 8. In addition to this, there are at least six or seven herds in which losses occurred from which the figures are not at hand. No one can consider inoculation to have been a success from this showing, when the losses among the inoculated herds of the state only reached 4 per cent. In other words, the inoculation of 3,000 hogs in 1891, instead of reducing the percentage of loss, increased it three-fold.

I have already written more than I intended, but in concluding I would remind you again that the failure of inoculation was sufficiently demonstrated in Bulletin No. 8 by incontestable evidence not contained in any of the correspondence to which you and Mr. Walker refer. This failure is demonstrated by the careful scientific tests made by the Bureau of Animal Industry; by the inoculations made in Nebraska in 1888, where over half of the animals operated upon afterwards died of cholera; by the complete failure to introduce inoculation as a private enterprise at Chicago; by the loss of half of the single inoculated hogs tested in the Peoria distilleries [see p. 217]; by the still more disastrous experience at Davenport [see p. 178]; by the communication of disease to the experimental hogs inoculated at Ottawa, Ill. [see index], and by the Nebraska inoculations of 1891. There can be no dispute about the facts in thesecases, and the attempts to explain away these facts have only made the weakness of the case more apparent.

If you considered it a duty to inform me at such length of the supposed errors in detail contained in the letters published in Bulletin No. 8, why do you not consider it a still more pressing duty to inform the farmers of Nebraska and of the country of the misleading statements which have been issued from the patho-biological laboratory in regard to inoculation? They have been misinformed, not only in regard to the details, but by broad statements, that inoculation was a great success, although it was proved a miserable failure. Many who have accepted these statements and acted upon them have lost their hogs by so doing, and many others are liable to meet with similar misfortunes from the same cause. In publicly sustaining your laboratory, in praising its work, in endeavoring to show that this department is wrong on this important question, you assume a very grave responsibility for the results.

Hoping that this letter will relieve your mind of doubts as to whether "the head of the department has really given the matter any thought at all," I am very respectfully, J. M. Rusk.

HON. CHAS. H. WALKER, OF SURPRISE, NEB., ANSWERS MR. RUSK.

Hon. Jerry Rusk, Secretary of Agriculture.

Dear Sir: I may owe you an apology for the part I have taken in exposing the false witnesses introduced in your Bulletin No. 8, in my letter to *The Nebraska State Journal* [see p. 144] to which you refer in a recent open letter to Chancellor Canfield. I did not intend to irritate you. I thought you would thank me for the information that you had been grossly imposed upon. In my innocence I thought you wanted the exact truth whether the testimony sustained your theory or not. I cannot but regret that it appears that I was mistaken. While it may make you a little impatient to have your attention called to these false statements, the hog growers, who are specially interested in this matter, are entitled to know the truth, let it come where it will.

In presenting a different experience from the one you have enjoyed, I do not mean to be disrespectful. I recognize your high position and do not wish to disturb your sensitive nature at the same time. Without intention, no doubt, it does seem to me that you have not done exactly square in this matter.. When you asked men (I mean your department, of course) to give you their experience, and they did it, it don't look right to suppress it and publish a different version from others not in possession of the facts unless from hearsay, even if it is more to your advantage in the discussion. It not only hurts the feelings of the persons thus treated, but it misleads and deceives those you are supposed to serve. As to the claim that mistakes have been made: Whatever others may have said I have never contended to the contrary. It is neither true nor wise to do so. Neither can I see how any reasonable person could expect such a thing, but I am unwilling that a matter of this importance shall suffer falsely or stand responsible for misfortunes arising from other causes.

You are exceedingly kind in excusing your witness because of a defective memory. The loss of memory is a great affliction and while you have excited my sympathy, it is still bothering me to understand how a man can forget so peculiarly as to think another had seventy-nine hogs inoculated and lost all but thirteen when he never had a hog inoculated in the world; but I know there are wonderful things transpiring in this eventful age.

You seem to think you have proven inoculation a failure by negative evidence. You think you have demonstrated by the failures you have made and because of them you urge that others who have met with success are deluded.

You do not seem to think that intelligent men that have spent their lives in raising hogs, and have anxiously watched them as they have lost them by the hundreds, are capable of passing on their own experience on this subject. With some of us, for years before we commenced the practice of inoculation, cholera was a regular visitor to our herds, while since that time we have been free from the scourge. If this is not the result of inoculation as you suggest, it is certainly a very happy coincidence. You have told us how you have been able to disprove the preventive powers of inoculation; by following you in our treatment of our hogs we ought to meet with the same results, but we have not. We have placed them in infected herds, the worst we could find, and have fed them on the viscera of hogs that had died with the cholera, without communicating the disease. And we have often repeated it. We have done it as honestly and thoroughly as our intelligence would permit from purely a selfish motive. We have not thought it would pay to deceive ourselves, and inasmuch as there is not a dollar in it if we were to deceive others, that motive, from a selfish or any other standpoint, could not be attributed to us. For these reasons we have tried not only to be honest with ourselves but to be as thorough in our investigations as we could, and when we found by experience we could put our infected pigs into infected herds, and feed them on the dead carcasses of hogs that died with cholera we were satisfied with the result. Experience extending over the last four years has warranted us in this conclusion.

If you had been a great sufferer from this disease to the extent that you had been kept poor by your losses would you think it a delusion if, upon adopting a practice recommended as a preventive, it should prove so in your experience? If it was a delusion would you not cling to it?

You say inoculation failed at Peoria, but by implication you admit that it did not fail with those that were twice inoculated. This is a destructive admission to make for your theory unless you claim that you have fathomed all the secrets controlling it, when you admit that any protection is received from the practice you have admitted too much—your case is lost.

If it can be done once it can be done again.

You say it was a failure at Davenport. You mean, if you wish to state the facts, that hogs could not be inoculated in the pens at Davenport without a great loss, for reasons that with experience became obvious, but you do not mean to say that hogs that had been inoculated took the cholera when exposed in these pens. If you do you mean to state that which is not true.

The enterprise was not abandoned because inoculated hogs could not be kept there, but because the operation could not be done successfully; that is, without too great a mortality on account of the food. This fact was thoroughly established with 200 inoculated taken there from this state.

You say that the experience in Nebraska in 1888 demonstrated the failure of inoculation, and I suppose you will continue to say so, no matter how conclusively it is shown that what you are pleased to call failures are simply the result of a natural outbreak of the disease that was raging in the neighborhood at the time and attacked herds that were not inoculated as well as those that were recently inoculated.

It would hardly seem necessary to contend with scientists that nature has but one law to govern the action of a certain poison. Still you claim to have found another. You have here a poison administered to several herds taken from the same bottle, and you claim as a scientist that while it did not even sicken one herd it killed 90 per cent of the other. You have been repeatedly asked for a precedent to help sustain your theory or any evidence to show that any poison is governed by two different laws that act haphazard or at pleasure, making itself deadly in one herd and harmless in the other. You rely on your table on page 11 of your bulletin to establish this. Mr. Sylvester's hogs were known to be sick when they were inoculated and dying, Mr. F. W. Luddon's hogs were never sick. Two other Luddon brothers occupied the farm across the road, one of them had his hogs inoculated and a large per cent of them sickened and died. The other had his quarantined in a pen and they were not inoculated, yet these also sickened and died in the same ratio. If it was inoculation that killed those that were inoculated what was it that killed those that were not, and if inoculation was so fatal to the one herd, why did it not even sicken the other Luddon's herd? You may be able by only telling a part of the story to convey the impression that inoculation was the cause of this mortality; try it by telling the whole story and see how it will work.

You claim a loss of 10 per cent of pigs inoculated. From that showing is it not better to lose 10 per cent of your pigs when they simply represent a sow's time for four or five months than to lose the 4 per cent you claim is the only loss of hogs not inoculated, with the corn they have consumed thrown in? But you do not tell us that men only inoculate that have places infected with cholera and then ask us to give you credit for 10 per cent loss on those infected grounds, but when you estimate the losses from cholera by natural infection you estimate upon all the hogs in the country, a very small per cent of which are kept on infected grounds, and yet you would have us believe you intend to be honest in your presentation.

Surprise, Neb. C. H. WALKER.

RESUMÉ.

Is the Secretary of Agriculture, Hon. J. M. Rusk, Guilty of Treason to his Official Trust or Not?

RESUME.

In his letter to Senator Paddock Mr. Rusk tries to make a point against me when he says, "He now turns to the people of Nebraska and frantically calls on them to sustain him by writing to their congressmen to 'stir them up,' he says, 'and let them know that they are your representatives and not the subjects of Jerry Rusk or his department." Such an accusation as that is supremely ridiculous regarding a man of my character, who is continually challenging public opinion on matters it unfortunately deems of far more importance to its spiritual welfare than the treasonable acts of the secretary of agriculture and his advisers.

In another part of this work I have printed two letters from the chief of the Bureau of Animal Industry begging for support from the agricultural press. Though written coolly enough, they demonstrate quite a "frantic" condition of the mind on the part of the department. It is well known that there is not and has not been one single first-class agricultural paper in the west on the side of the department for some years, as it is also equally well known that bribery, in the form of employment, has been given to about the only paper which does support its opinions. These facts speak for themselves.

Has not the utter fallacy of every claim made by Mr. Rusk been demonstrated time and again, and even now by the fact that not one responsible western agricultural paper has ever noticed such appeals for aid from these politicians, who are slowly drowning in the slough of despond, filthy and corrupt with their own errors and misstatements? The editors of these papers cannot afford to stultify themselves and defy the intelligence of their readers, as the secretary of agriculture and his advisers do. Where, in the last five years, can be found an editorial supporting work which the chief of the Bureau of Animal Industry claims to be "the careful and comprehensive researches" of that department? All over the west the farmers have accepted these researches with that sullen indifference which characterizes the reception of the fair promises of many of their deceiving employers by the wage earners. I beg "frantically" for support in a position nothing but duty to the state of Nebraska causes me to occupy!

Mr. Rusk defies the verdict, not only of this state, but the great agricultural press of the west, when he supports himself on the say of one self-looking-out-for-employe, who has nothing he can show in his fourteen years' work for his department against this common verdict, that nothing of value has emanated from his researches by the intelligent among the people and the scientific world. It is easy to deceive one man when that man is so ignorant of the necessary details of the work of his department as to be totally unfitted for the position. It is a disgrace to the intelligence of the people, a threatening danger to the cause of good government, to fill a position requiring that broad education, that intensity of enthusiasm which will enable the occupant to live, sleep, and eat in earnest and intelligent sympathy with the varied interests of the American farmer with any man so ignorant and incompetent that he can be so easily imposed on as the secretary of agriculture has been, merely because such a man is a sort of additional wheel in the political machine, and has influence enough to add an indefinite number of corrupt spokes to strengthen the machine in its daily work of grinding out the chains of slavery for an indulgent and easily imposed on people. Does the secretary of agriculture know of any people so self-abnegating of all personal interest that they are willing to hold a man in office at my salary and spend ten thousand a year merely out of personal liking for him, or merely because he "frantically" cries for aid in his dire distress? Human nature is not built that way according to my experience.

While I have perhaps two friends in Nebraska, Mr. Gere, editor of the *State Journal*, and Mr. Reed, of *Western Resources*, who would deny themselves much to aid me were I in distress, men who love me because in doing so there is something in me which stimulates their own self-love (no man loves another, or woman either, save for his own pleasure; no man is friendly to another save that it gives him happiness; no man hath greater love for another than that he gives his life for him, because nothing else yields him so much happiness), still the mass of people who desire my presence in Nebraska do it on the same principle, for self in another form, *because my researches have not only done them some good, but promise to do more, without one word from me in their support.* Mr. Rusk denies to these men the attribute of common intelligence, that honesty to themselves which enables them to judge of what is best for their own interests. Mr. Rusk refuses to listen to these men, but has an open ear for a man who is absolutely without public confidence or support, no matter how frantically he may plead for them. The secretary of agriculture knows this to be so, but, because he thinks, or has been led to think, that in some way my exposures of the rottenness of the researches of the Bureau of Animal Industry encroach on his dignity, Mr. Rusk refuses to listen to the verdict of the people, and publicly defies them. "There are none so blind as those who wont see." There are none so easily misled as those so ignorant as to be unable to appreciate the truth when squarely presented to them. No greater danger threatens us than the boldness with which politicians continually defy the intelligence of the people. As a nation we are a generation of slaves, bowing meekly before the political juggernaut. Thankfully, portions of the nation are waking to a necessity of manhood. The "people's movement," in its various forms of farmers' alliance and labor organizations, is the most promising sign of the times.

LET US SUM UP THE EVIDENCE AND SEE WHETHER THE SECRETARY OF AGRICULTURE IS GUILTY OR NOT GUILTY OF TREASON TO THE RESPONSIBILITIES OF HIS HIGH OFFICE?

As admitted by Mr. Rusk in his letter to Senator Paddock when he entered upon the duties as secretary of agriculture, he soon found that a strong dispute existed as to the results published in the reports of that department and my own, and that his predecessor had appointed a "Board of Inquiry" of "eminent scientists" to investigate that matter.

In May, 1889, one of that board, Dr. Bolton, sent in a report signed by himself, which fact of itself, is evidence enough that he did not fully agree with what he must have known was to be the findings of the other members of that board, and that he would not trust them, or else he would have handed his report to them, to be handed with theirs by his colleagues on the board to the secretary of agriculture. This report of Dr. Bolton did not conform with that of the two other "eminent scientists," who handed in their report August 1, 1889, as shown, not by words of mine, but by the verdict of the agricultural press and some of the most "eminent scientists" in the true sense, among the investigators of the world. It is self-evident that the secretary of agriculture must then have had the two reports in his hands August 1, 1889. What explanation then can he give the people of this country, whose servant he is, for "a public office is or should be a public trust," for authorizing the publication of the second report received August 1, 1889, and withholding the first report received May, 1889, until some time afterwards? Is it not self-evident that the secretary of agriculture gave his consent to this infamous procedure in order to do me personal injury, and to deceive the press and farmers of the country? What is that but treason? Can there be any greater guilt on the part of a public officer? That it was considered to be a deception of the public by the editors of the agricultural press has been demonstrated in the previous publication of their editorials written at the time. In those editorials the editors of such sterling papers as the *Breeders' Gazette*, the *Farmers' Review*, *The Homestead*, and *Western Resources*, as well as others, asked the Department of Agriculture to explain the self-evident contradictions between the "fly sheet" majority report sent to them first and the later published minority report of Dr. Bolton? By entirely ignoring those questions, by ignoring the universally appreciable fact that the reports of that board of inquiry fell flat, does not Mr. Rusk admit his guilt of malfeasance in his office, and of his willful attempt to deceive the public? Is that being guilty of treason or not?

In his letter to Senator Paddock the secretary of agriculture betrays a most intimate acquaintance with the various publications from myself, and necessarily with the editorials of Hon. Charles H. Gere, also president of the board of regents at the time. In Bulletin No. 8 quotations are made from my Chicago "Pamphlet No. 3." In the columns of the *State Journal* as well as in that pamphlet appeared the following self-incriminating letter from one of the "eminent scientists" of that "board of inquiry concerning swine diseases," which the reader must pardon me for introducing again:

Under the heading, "The Report Was Fixed," the editor of *The Nebraska State Journal*, and who is also president of the board of regents of the State University, published the following in a late issue:

"The suggestions made by *The Journal* after the report of the swine disease commission was published last summer, that some undue influence was exercised on the members thereof to induce them to suppress all that they had really found out by their protracted investigation, is pretty well borne out by a letter addressed by a member of the commission to Dr. Detmers, of Columbus, on the day before the report was published. The letter was as follows:

"'AGRICULTURAL EXPERIMENT STATION,
"'UNIVERSITY OF ILLINOIS,
"'CHAMPAIGN, ILL., August 13, 1889.

"'MY DEAR DR. DETMERS: I suppose you have seen by the papers that the report of the "board of inquiry concerning diseases of swine" has been presented to the Washington authorities, and I suppose, too, that you are disappointed at least in my signing the clause in respect to your early work. Yet I know fully that you would not feel hard if you knew what I did do and what I tried to do and failed. I could not help knowing that the germ you worked with in 1878–80 was the same as that now held by you to be the cause of the disease you call swine plague. I knew this without examination this last winter. I have not said you are not entitled to priority in this, neither has the board. After a little, when the time seems to be proper, I want to say as an individual what I know and what should have been said in this report.

"'While I could not help doing what was done, I was far from satisfied.

"'As to there being two diseases with different germs we were all agreed; but not satisfied as to the commonness of occurrence of the one called swine plague by the Washington folks. We found one outbreak away from the east where it had only been found near Washington and Baltimore. It is not probable that it is a prevalent and serious malady, though it seems bad enough when once introduced to a herd. The most that troubled me about it was whether or not it was a peculiar variety of the common germ. But it has so many points of difference and holds them so tenaciously in all process of culture that it seems to me impossible to consider them as one thing in any sense. This one new outbreak was in Kentucky. I wanted to go back and further study the disease there, but neither had opportunity or permission. Hoping to see you soon, I am faithfully yours,

"'T. J. BURRILL.'

"The language of the letter is positive proof that the commission was making a certain sort of report under compulsion, and as the commission was acting under the authority of congress and was paid by an appropriation of $30,000 made by congress, *The Journal* calls for a congressional investigation of that report which has cost the country such a smart sum."

With the evidence that we have, that Mr. Rusk cannot help but know of the above letter, and that to any honest secretary of agriculture determined to do his duty, the letter of Mr. Burrill shows self-evidently that he had not done as an honest man should, was it not the duty of Mr. Rusk to ask of Mr. Burrill for that evidence which he promises "*after a little, when the time seems to be proper, I want to say as an individual what I know, and what should have been said in this report*"? How much time does Mr. Rusk require to do his duty and force a deceiving official to make explanations promised August 13, 1889?

In the columns of the *Nebraska State Journal*, and also at different times from me, appeared the following quotations; the first from an article by Dr. Frosch, assistant in Koch's laboratory in Berlin, who, by Koch's orders, passed in review the literature of both sides of the controversy in this country and then compared the results here by personal investigations and decided in favor of the Nebraska investigations; the second, by another careful reviewer, who went over the whole literature in most critical manner and published the results in the *Journal of Comparative Medicine.* Dr. Frosch says:

The fact stands for me confirmed that in the five years that have passed since the second plague was first announced not one single case of an independent appearance of Salmon's swine plague has been reported which can be said to be free from objections.

The mistaken conception of Salmon's place in the swine plague investigations must be laid to the fact that the publications, both on swine plague and hog cholera, have his name and as such have been quoted in the literature.

From the present publication of Smith's, however, which could not be seen in reading the reports of the Bureau of Animal Industry, it is evident that Salmon was not the discoverer of either the hog cholera germ or that of the swine plague, so now we know the true condition of things in that regard.

The reviewer in the *Journal of Comparative Medicine*, says:

We have compared the extracts cited by Dr. Billings from the reports of the Department of Agriculture, line by line and word for word; we have carefully examined the plates accompanying these reports to which he refers, and we have also read the entire context from which he quotes, in order to avoid any possible bias which may follow from reading disconnected sentences. As a result we are enabled to say that all of Dr. Billings' charges, sweeping and severe as they are, are true and just, and we cannot understand how the committee appointed to investigate the special question involved could fail to make the same examination, to arrive at the same result, and to publish that result in calm, judicious language, not as a matter of justice to Dr. Billings, or of criticism of Dr. Salmon, but as a positive duty to science. Regarded from this point of view, the report of that committee is the most disappointing document of the kind we have ever seen. This we may say without intimating, as Dr. Billings does, that its authors were under the control of the "Bureaucratic Whip."

Mr. Rusk cannot claim ignorance of these things, and yet he passes them by and utterly ignores them. Would such conduct not be considered treason to a public trust in an official in any other government and by any other than the blindly indifferent American people? The statistics and other evidence furnished by Mr. Rusk's correspondents in "Bulletin No. 8," on Inoculation, have been given in this work, and it has been conclusively demonstrated that *"by the authority of the secretary of agriculture"* not only was self-evidently false testimony given by his correspondents asserting losses by inoculation on the part of men who never inoculated a hog, and in districts where no inoculation has ever been done, and that some of this evidence is even contradicted in the bulletin itself, where in the table given F. W. Luddon is said to have lost "none," and yet several correspondents say he lost nearly all, and of the case of his brother Charles who never inoculated a hog, and yet is said to have lost equally from its effects, but beyond this it has positively been shown that many letters (we have not published all we have) favorable to our work which were written to Mr. Rusk must have been received and were purposely suppressed in that bulletin.

For what is the secretary of agriculture appointed? Is it to lay the whole unvarnished truth before the farmers of the country, or is it to deceive them by the publications of absolute falsehoods, erroneous statements, and misleading testimony, and by withholding true evidence from men whose repulation is beyond question? Is Mr. Rusk guilty of treason to the trust imposed on him, or not?

At Ottawa, Ill., Mr. Rusk first told the farmers that he could do nothing for them for lack of funds, but when he heard that we were going to try to aid them in their distress he found abundant means to send agents there to create distrust in our endeavors, and to nullify their effects. Acting by his authority, these agents proposed a test and that the farmers select a committee of referees in whose hands the whole matter was to be left. Acting under his authority, instead of immediately leaving, as an honest disputant in such a question should, after inoculating his bunch of hogs according to the bureau method, and as Mr. Cadwell did, no matter how he did it, Mr. Rusk's agent remained on the ground and took the whole matter out of the hands of the referees, issuing reports at his pleasure, or by command from Washington, and finally, I am informed, writing the report which the contemptible committee of referees signed. It matters nothing to the question which way the results went. I ask is that an honorable and dignified course for an official in Mr. Rusk's high position to sanction? Would any set of men in the ordinary business of life so conduct themselves? Would any court of justice in the world sanction such a procedure for a moment? Is not that one of the highest punishable crimes, "tampering with the jury"?

THE SECRETARY OF AGRICULTURE INTERFERES WITH THE CONSTITUTIONAL RIGHTS OF PRIVATE CITIZENSHIP.

The misdemeanors on the part of secretary of agriculture, or those having his sanction and authorization, previously considered, have been largely in reference to the public, especially the farmers and swine breeders of the west. I must now beg the indulgence of my readers for drawing attention to an act which bears more directly on myself, though fortunately for me its repetition in that "Farmers' Bulletin No. 8" renders it a matter of equal importance to the public. Had it not been for this I should have passed it by, as not worthy of notice. As has been repeatedly noticed, this Bulletin No. 8 bears on its cover the words "published by the authority of the secretary of agriculture." I now am going to call attention to the most disgraceful statement in the whole bulletin, *a statement which has no place whatever in that bulletin; a statement which has no connection with inoculation as performed in Nebraska; a statement which lays the secretary of agriculture open to prosecution in the courts of the land; a statement made only to mislead the public and to create a prejudice against inoculation; a statement which, stronger than any other made in that bulletin, shows its malicious and villainous character, and proves beyond question that the Department of Agriculture knows that inoculation has been successfully done; that it can be done, and that it is going to be made successful, and that it will then kill its opponents and justify the battle which I have made not only for inoculation, because I know it will succeed in the future infinitely better than in the past, but for justice and truth towards the farmers of the United States.* Inoculation will succeed, and with it must come honesty in the public service as represented by the Agricultural Department. Here is the statement:

THE FINANCIAL ASPECT OF INOCULATION.

It is very apparent from the facts presented in this bulletin that inoculation is a very dangerous operation, and that the protection from it is, at best, uncertain, and in many cases entirely wanting. With these incontestable conclusions in mind, we will give some figures on the losses from swine disease and the cost of inoculation. Two years ago the following statement was made:

According to the estimates of the statistical division there are about 50,300,000 hogs in the United States. The inoculation of these at 50 cents per head would cost $25,150,000. The total loss from disease during the year 1888 was 3,105,000 hogs, at an average value of $5.79 each. This would make the total loss of swine from all diseases $17,980,000.

In order to estimate the loss from hog cholera we must deduct from this sum the losses from ordinary diseases, such as animal parasites, exposure, overcrowding, and improper feeding, which are always acting and do not produce epizootic diseases. Those losses were estimated by the statistician of the department in 1886 to be about 4 per cent of the total number of hogs, but as this may be considered rather a large estimate we will, in our calculation, take 3 per cent as the average loss from such causes. This would amount in 1888 to 1,509,000 animals, valued at $8,737,000, and deducting from this the total loss of swine we have remaining $9,243,000 as the losses from epizootic swine diseases. In the present condition of our knowledge we must admit that there are at least two entirely distinct epizootic diseases of hogs, which have been referred to in the reports of this bureau as hog cholera and swine plague. The exact proportion of the loss caused by each of these diseases is at present unknown, but if we admit for the purposes of this calculation that but one-third of the loss is caused by swine plague we have remaining a loss of but $6,163,000 for the year 1888, which can be attributed to hog cholera. To prevent this disease by inoculation, as we have just seen, requires the expenditure in cash of $25,150,000, or more than four times the amount of the actual losses. In addition to this expenditure there should be counted the time required by the farmer in handling the hogs at the time of the operation and in giving them such precautionary care and in practicing such disinfection as is required to make this operation at all successful.

We should reach the same conclusion if, instead of estimating the loss and expense for the whole of the United States, we should take a single hog-raising state, as for example, the state of Illinois. According to the statistician's estimate there are 5,275,000 hogs in Illinois, and to protect these by inoculation would cost $2,637,000. In the year 1888 the total losses of hogs in that state from all diseases was about 316,500, with an averagevalue of $7.45 each, which would make the loss for that year $2,359,925. Deduct a loss of 3 per cent of all the hogs in the state as caused by ordinary diseases, and we find that this would amount to 158,250 hogs, worth $1,178,962. Deducting the losses caused by ordinary diseases from the total losses from all diseases and we have $1,180,963 left to represent the loss from both hog cholera and swine plague. Take from this one-third, to represent the loss from swine plague, and we have the remaining, as the loss from hog cholera, about the sum of $800,000. To prevent this loss by inoculation, as we have seen, would require $2,637,000, or more than three times the sum to be saved.

In the above calculations we were considering inoculation when practiced as a private enterprise, with a charge of 50 cents per head for the operation. It has since been proposed that the virus and instruments should be supplied by the state experiment stations and that the farmers should perform the operation themselves. This would no doubt reduce the cost of inoculation to 25 cents a head for the time and trouble involved in the operation, the expressage on the instruments and virus, and the precautions necessary to prevent the spread of the disease to other herds. To this we must now add the loss following the operation when performed on healthy herds. This we have just seen has been with 2,643 animals inoculated the last year, and with every precaution that could be adopted, over 10 per cent. If the hogs average $5 per head in value this would be an additional expense of 50 cents per head for each inoculated animal.

What truth is there in this statement?

It has been proposed that the virus and instruments should be supplied by the state experiment stations and that the farmers should perform the operation themselves. *This would no doubt reduce the cost of inoculation to 25 cents a head for the time and trouble involved in the operation and the expressage on instruments and virus and the precautions necessary to prevent the spread of the disease to other herds.*

We have nothing to do with other experiment stations, but here is the cost to Nebraska farmers:

While the virus and necessary implements will be supplied to any farmer or breeder in Nebraska free of charge, except that such farmer or breeder must pay the express charges on the same from Lincoln, and for the return of the implements to Lincoln, *still any person who neglects to return said implements will be shut off, and in future shut off from the privilege of obtaining virus from this laboratory.*

Now, by what manner of figuring inoculation can be made to cost "25 cents a head" under those conditions, unless limited to two hogs, is more than I can see? We can send virus for 1,000 full grown hogs anywhere in the state for 50 cents, expressage both ways, and for $1.00 for 10,000, so the deceptive fallacy of such an argument is at once apparent. As to expenses for the "precautions necessary to prevent the spread of the disease to other herds," they do not exist, because, as we only inoculate herds already on dangerous ground, the danger existed before the inoculation, which is the reason farmers inoculate, and as no precautions can be taken beforehand, other than have been mentioned, of keeping the hogs on those grounds and all other people and things off them even before inoculation, it cannot be seen how the expense is increased an iota by inoculation. As to losses, stunting, and dangers, our correspondents, whom Mr. Rusk also received letters from, have shown none such exist with the virus sent from here.

To whom then is that statement directed? What cause was there to print it in that bulletin? It certainly has no relation to the work done in Nebraska. Why then was it printed? If not printed to convey a false impression, and as the last resort of desperate men to prevent the successful spread of inoculation among the farmers of the country by endeavoring to frighten them on account of the expense, then for what purpose was it printed? It cannot have been intended for me now, for in no way possible can I make one cent, nor do I endeavor to make a cent, even where I could, out of inoculation in other states, or even by writing letters of advice to farmers in other states. It can only be directed then at my late partner, Mr. George A. Seaverns, of Chicago, a gentleman to whom the farmers of this country owe an everlasting debt, for had it not been for him inoculation would have been a thing of the past now. It cannot even be directed against Mr. Seaverns, for long ago he withdrew all advertisements and gave up inoculation because it was useless for him to go on with his government doing all it could to injure him. Nevertheless, the secretary of agriculture should be financially and officially responsible for the publication of such stuff as the above, for the reason that it is the same malignant assault which he authorized to injure the business of myself and Mr. Seaverns when we were together. In Bulletin No. 8 it says that the endeavor to make a business out of inoculation was a failure. How could it have been anything else with the whole power of the Department of Agriculture opposing it with such stuff as that? The first attack of that kind was made before the National Swine Breeders' Association in Chicago, November, 1889, and published in the reports of that meeting and somewhat in the daily press. The thing was repeated at the meeting of the Kansas State Board of Agriculture a few months later, and published in its report; and again in a special bulletin of the Department of Agriculture in 1890 sent broadcast over the country.

Notwithstanding the intrigues of the Department of Agriculture had forced me to leave Nebraska in 1889, out of self-respect, the secretary of agriculture still pursued me, as a private citizen, and did his utmost to ruin me financially, utterly regardless of the effects on my family, and his malignant interference caused Mr. Seaverns, an entirely innocent man, and a quiet business citizen of the country, an earnest republican too, at that, to lose some fifty thousand dollars.

Remember, reader, Mr. Rusk did not publish in any of these statements that inoculation was a fraud as a fact, nor does he now; he did not say then that it has not been successfully performed many times, nor does he now; he simply told the farmers then, as he falsely intimates to them now, *that it is too expensive.* Is that any of Mr. Rusk's business? That is a question for the farmer alone to decide. We will not discuss it. But suppose Mr. Seaverns had still continued it; suppose he desires to resume it, does the constitution of the United States give the secretary of agriculture a right to injure a citizen's business by telling those likely to be his patrons, *that what he offers them is too expensive?* Where do we live? Is this the land of Paine and Washington, Hancock and Adams, Jefferson and Randolph, established in the hopes of making an ideally free people, or is it inquisitional Russia with its infamous czar? Has not Mr. Rusk broken the constitution in letter and in spirit? If so, then I declare the secretary of agriculture to be the greatest traitor, the most malignantly false official that has ever disgraced this government since the fourth of July was made gloriously immortal in the sublime words of the Declaration of Independence.

Changing the wording, but in the spirit which animated Patrick Henry, the American people of to-day should say, *"give us justice or give us death."* In the same way, and for the people of the United

States to consider, I would change a passage of Secretary Rusk's letter to Senator Paddock to read:

> If the people of the United States wish to continue this kind of a secretary in such a conspicuous position, paying him $8,000 a year and allowing him to spend the public funds to injure their live stock interests, I suppose they have the power to do so, but they cannot expect to shift upon other parties the discredit and disgrace which the misdoings and neglect of the Department of Agriculture bring on the fair name of the country.

CONCLUSION.

The Causes Which Have Led to the Conditions Here Portrayed in American Politics.

CONCLUSION.

"*When precedents fail to assist us we must return to the first principles of things for information, and think as if we were the first man that ever thought.*"

"*Back to the first plain truths of nature, friend, and begin anew.*"—(Thomas Paine.)

In the foregoing pages have been chronicled a series of facts of the most disgraceful character in connection with purity in the public service of the government of the United States. Wherever we have an effect we at once know that there must be some cause or causes behind it. Nothing happens without cause. Among these causes have been clearly demonstrated:

First—Erroneous work and false publication in the reports of the Department of Agriculture having reference to the work of the Bureau of Animal Industry.

Second—Cowardice, due to fear that such erroneous and false work would find exposure.

Third—Jealousy. The evident desire to make the public think that all the work on which it should depend in these lines emanated from the National Agricultural Department.

Fourth—Failure to do anything after more than ten years' trial in the public service.

Fifth—Fear of the effect of the Nebraska investigation on the minds of the farmers in the coming election.

"A DEMOCRAT FROM DEMOCRATVILLE."

Political persecution of a life-destroying character, while not of the active guillotine type, is nevertheless a fact in this country. The right of independent citizenship is denied to every person in the public service by the party in power, unless the person holding such position is the subservient tool of the temporary rulers. A spy system is kept up in order to watch the movements and catch the words of minor public officials. Some of the higher need to be watched also by their superiors. It is a well known fact that the delegates at Minneapolis received numerous telegrams from Washington to "*support Rusk for first place if Harrison won't work,*" or words to that effect. There is intrigue everywhere and honesty to one another is made painfully pregnant by its absence.

The spy system which has been kept up on the work of this laboratory, especially during the last year, notwithstanding the poverty of Mr. Rusk's department, has been something unheard of, where everything was open to the public. Only the other day a gentleman in whom I should be able to trust implicity told me that he saw a letter from the secretary of agriculture in which it said that he, the secretary, "had it from a stenographer who took down the words as spoken by me that I said I was a "*Democrat from Democratville,*" which is about the only true accusation which has been made against me. It is no new thing, the same charge was made before the investigating committee of the Nebraska legislature of 1888–9. The same charge was one of the causes of the "Bill for Remuneration" being put into the legislature to demand pay for the hogs claimed to have been killed by inoculation at Surprise, Neb., in 1888. Mr. Rusk refers to that bill in his letter to Chancellor Canfield and endeavors to draw support from it, that the hogs mentioned were indeed killed as he claims. Mr. Rusk does not tell his readers that the legislature of Nebraska, even though heated to a red-hot temperature, did not believe that charge to be true; that it did not consider that bill even in committee, so far as the public knew of it; that it never was brought up by the "investigating committee," though challenged to do so in the following "open letter" published by me in the Lincoln *Daily Call* of February 14, 1889:

I see by the columns of the *Call* of last evening that certain ignorant, demented, and unfortunate people, who have been led to think that they lost certain hogs by means of inoculation through the instigation of a small gang of political dead-beats who live on the public pap, and a certain pole-cat whose offensive aroma still pollutes the pure air of Nebraska, have presented a bill for remuneration for their own ignorance and laziness. It is my desire to stir up this pool of filth to its dregs, and now I hope every farmer in the legislature will push "Hill's hog cholera bill" to an abrupt conclusion. The remark that "they did permit him (me) at his (my) request to inoculate their hogs," on the part of said owners, is an infernal lie. The assertion that one single hog of the 614 inoculated at that time and place, with the same material, died from the effects of inoculation, is also an infernal lie, and the persons who have instigated this move and those who have followed at their beck and call, and every one who supports it, will be guilty of an attempt of as barefaced a robbery of the public funds as was ever attempted, though small in amount.

These words were bold enough to have stirred up a hornet's nest of action in a mad republican legislature, backed by a frantic governor, had there been an iota of "sand" in the whole body. The trouble was they knew full well that they did not have an iota of fact to stand on. They knew that to act on that bill as desired would bring down on them every intelligent farmer and breeder in the state. Though a "democrat," the machine could not decapitate me. It cannot do it now. My strongest supporters are all republicans. It is an encouraging sign of the "good times coming," to see the intelligent men of this state stamping the verdict of the machine to pieces and standing up for themselves. Does the secretary of agriculture assume that he will do anything but increase the contempt and distrust which the breeders of Nebraska already have for him when they read in his letter to Mr. Canfield that "it may suit the purposes of the half dozen persons who are interested in sustaining inoculation to draw the attention of the farmers from the facts which prove it to be a humbug, by assuming such unbounded indignation over errors in a few letters which could be excluded without affecting in the least the general conclusions of the bulletin. But unless I am greatly mistaken as to the intelligence of Nebraska farmers, this plan of throwing dust in their eyes will not succeed"?

It will be noticed Mr. Rusk admits the "errors which could be excluded," but were not. Why not? Certainly the secretary of agriculture must have "lost his head." He cannot know the facts, notwithstanding the diligence with which he claims to have read the Nebraska papers. Will he kindly mention the "half dozen persons" by name who have such direct and personal interest in inoculation, that they desire to "humbug the farmers"?

To be sure there are just six members in the Board of Regents of the State University, but they have "washed their hands" of this matter and do as requested by the breeders of the state.

Is it the State Board of Agriculture and county agricultural societies of the state; is it the Nebraska Improved Live Stock Breeders' Association, composed of representive men directly interested in breeding all kinds of stock; is it the Shorthorn Breeders', the Hereford Breeders', the Draft Horse Breeders', the Trotting Horse Breeders', the Swine Breeders', or the Nebraska Dairymen's Association, to whom the secretary of agriculture refers as so low and despicable that they support a man in position for some perfidious and personal purpose?

Let us go slowly in this matter!

Something must be very rotten in Washington, or in Nebraska, when the secretary of agriculture dares to assert so many of the most responsible and respectable men of a state, every one of whom are farmers, are humbugging their brother farmers for some personal reason. No public official since our government was formed ever had such barefaced audacity as to accuse such a representative body of men as the members of the associations mentioned, of being either idiots, capable of being easily fooled by such an extremely clever and unprincipled fiend as the writer is asserted to be, or rascals capable of fooling their co-citizens for personal reasons.

As in every other state in this union, the men who support and control this work in Nebraska not only represent the most intelligent farmers in the community but the most public spirited. They have formed themselves into the associations named not only for mutual benefit, but because unitedly they can better develop the particular interest named, which is of vast importance to the state. The great majority of those men are republicans, but they never asked whether I was a "democrat from Democratville" or not, because they all know, and I know I dare assert it, *that whatever my politics may be, that above all I have been and am as true as the magnet to its pole to their respective interests, the live stock of Nebraska, and that means as well of the United States?* They know that it is none of their business what my politics are. They know and I know that it is none of Mr. Rusk's. They know and I know, and if Mr. Rusk does not know it, the quicker he learns it the more creditable will it be to his intelligence, that the only questions with which they have to do are: Is he competent to do the work we entrust to him? Is he honestly devoted to our interests? The secretary of agriculture denies both. He is but one man, and a totally incompetent one at that. Not only does he not know enough to judge, but so biased and unscrupulously dishonest is he that he refuses to see the truth and publish it when before him; so dishonestly partisan is he, that he tramples the declaration of independence under foot and sets the constitution at naught, because the writer happens to be a "democrat from Democratville."

There is too much liberty in the United States. Liberty which countenances abuse is the mother of anarchy, and yet it was for liberty which our fathers fought. Liberty, untempered by intelligent justice, is the entire cause of our social and political rottenness to-day. Let us close with a short consideration of this subject.

"WHERE LIBERTY IS NOT THAT IS MY COUNTRY."

Such were the words of Thomas Paine to Benjamin Franklin. Such was the spirit which animated uprising humanity at the end of the last century. Such was the spirit which inspired the "rights of man"; which filled the "crises" with that power that brought the American revolution to a successful ending and made the French revolution disgraceful by the imprisonment of the greatest defender of human rights the world has ever known. Unlike the American, the French revolution was a demand for liberty untempered by intelligent justice. The "liberty" then demanded was for self-government. It was a revolution against monarchial oppression. The spirit ruling in America was purer than in France. There was more self-abnegation and greater public devotion. We had no Robespierre to sully our country with bloody deeds of injustice. Public service has never yet been honored with the devotion of such a body of men as the "Fathers of the American Revolution" in any country. No other country has ever been in the same condition America was then. The strong men of the world, the liberty loving brains of humanity, had gradually centralized on the virgin soil of the United States. "Give us liberty or give us death" was their watchword. Their action gave birth to a nation with the slogan cry of "liberty." They little realized that nature knows no such thing as liberty. Nature is one moving example of subjugation to law. The "spirit of '76" was justified in demanding liberty from kingly rule. The spirit of to-day demands justice to all in the application of the laws. The sons are revolting against the unintentional liberty planted by their fathers.

The liberty of '76 has led to the anarchy of the present century.

The anarchists proper are not the dangerous element in society to-day. They are the logical result of the spirit of 1776. The truly dangerous elements are the "survivors in the struggle for existence"; the successful ones who, having strength to assume liberty for themselves, deny justice to every one else. The fathers failed in that "they did not go back to nature and study first principles." Had they done so, they would have seen that justice was greater than liberty; that rights were of human origin, forced by the might of concentrated mass power, concessions made necessary from those who had assumed the liberty unto themselves to do as they pleased. Our government is dangerously approaching a condition of anarchy. Our law-makers have less respect for the constitution and the laws than the people, whose servants they should be. Both the constitution and the laws are trampled boldly underfoot. The rich individualist defies them in his relations to the wage earners. The massed individualism of the trades unions defies them in its despotism toward free labor. The boss politician defies them in his attitude towards the servants of the people in the public service. Parties defy them in demanding the servile and unintelligent allegiance of the people. Crime finds its excuse and support in the defiance of the laws of the land by the very authorities created by the people to execute justice. Justice is trampled underfoot on all sides. Rule is based on precedents, rather than an intelligent knowledge of first principles. Blackstone, the British code, and Roman law are based too much on the precedents of ignorance, and lack almost entirely in a foundation on natural principles. Law is too artificial to be just. We must "back to the first principles of nature and begin again." Science was not sufficiently developed to permit the fathers to do this. Kings were too prodigous monsters in their eyes for them to see that the very liberty which they demanded and thought they had established would lead to a disrespect of law and become a threatening danger to the country a hundred years later. Little did they anticipate a nation of over sixty millions of people, one-third at least almost unable to survive in the turmoil of the busy social struggle which has come upon us. The country was to be the "land of the free and the home of the brave," but they did not reckon with the hordes of ignorant foreigners and semi-intelligent children that would be developed, all uncomfortable, all excited, and demanding that "liberty" so freely promised them and yet which can never be until the human race learns so to breed, itself that their kind will no longer be produced. The liberty which the fathers fought for was possible. That which they transmitted to their children is a rank impossibility. Only the extra-intelligent and profoundly educated man can have that liberty. Only the man so bred and educated that he is law unto himself; so just that he requires no law between man and man, can be trusted with supreme freedom. Such people are in the vast minority. Hence, the battle cry of to-day and the future must be *More education and profounder justice.*

THE DEMOCRACY OF TO-DAY.

We read and hear of the "democracy of America." Where is it? Is it to be found in the democratic party, with its Tammany machine, its saloon "inflooence," and-bold faced defiance of intelligent and thinking men? Is it to be found in the republican party, with its machine and its bosses; its undemocratic protection tariff, which robs the many for the benefit of the few; in its narrow nationalism, which again takes up the old cry of ignorant assumption, "America for Americans," in place of the nobler principle "*the world for humanity*"? No! the "democracy of America" is in the intelligent, the broadly educated, the truly humane instead of the exorbitantly selfish among the people, irrespective of party. The "democratic party," as such, with its bold defiance of the true democracy, with its corruption and corrupting influences, with its despotic defiance of the "mugwump" whom it sanctions because it cannot behead, is as false to every true democratic principle as its great opponent. A democracy cannot be a party. The sooner this lesson is learned the better. There can and must be a national democracy representing one demand only—profound justice, which should easily find one man pre-eminent above all others, as I think Cleveland is, in that one general direction; but there must also be special interests, each democratic, demanding justice, yet willing to deal justly, which should have their selected and special representatives.

"Conservatism," so called, which is egotistical individualism, cannot well form a party either, save in the same general way. Vested capital should be represented only by those especially competent to understand the necessities of a special interest. People engaged in a common livelihood all have a common interest. We want interest representation instead of partyism; communal-interest representation for the various and diversified interests of the country; minor democracies; and a president representing the just principle of intelligent justice of the great democracy of the country. In one sense the individualism of citizens, towns, and states must be lost in the welfare of the people as a whole; while nationalism must be tempered by that broad internationalism, with its world's congress, that armies and navies, tariffs and fortresses may disappear, and "peace on earth and good will to men," the result of even-tempered justice, may rule the humanities. This is no idle dream. Every student of evolution sees the inevitable coming of the social revolution which will make these things possible.

THE LIBERTY OF THE PAST CENTURY IS THE GREATEST OBSTRUCTION TO THE ADVANCE OF JUSTICE IN THIS.

"We hold these truths to be self-evident, that all men are created equal; that they are endowed by their Creator with certain inherent and inalienable rights; that among these are life, liberty, and the pursuit of happiness."

Such are the words of the "Declaration of Independence," or in other words, "liberty and equal rights." The theoretic purpose of the French revolution was expressed in the words, "Liberty, Equality, Fraternity."

Many people are so constituted that for any one to question the truth and correctness of either of the above expressions of the "spirit of '76" is considered sacrilege. Yet while from the spirit of a broad and philanthrophic humanity they express beautiful sentiments, they are both equally false to nature, and hence, have equally been and still are one of the principal causes of the coming social "revolution." It is to be hoped it will prove a peaceful evolution.

The well-known social conflict briefly referred to above, between capitalists and wage-earners, between the agricultural classes and capitalists, and between organized and free labor, are the natural results of this unrestricted freedom idea of the last century. Nowhere is it more pregnant than in this country. This "free and natural-rights" idea as naturally leads to the uttermost "go as you please" individualism, with all possible disregard of others, as is possible under our social conditions. The so-called capitalists represent the comfortable survivors in this great social struggle for existence. The wage-earners and the agriculturalist as truly represent those having less ability, who, under natural conditions, would live a desperate struggle and miserably perish in the end, as they have a tendency to now. Only by organizing and presenting a massed front to the superior abilities of the capitalistic class are they enabled to hold their own at all. Nature is not an intelligence, and has no purpose. Things have resulted from the action of fixed laws. They have not been created.

There is a tremendous word in what we call natural economy. It is "can." Our fathers failed to recognize its importance, though they practically applied it. Only those who have "inherent and inalienable rights to life, liberty, and the pursuit of happiness," who have the individual might, CAN force these things from nature. Might is right in nature. There is no other. Those who cannot force life and liberty from their surroundings certainly cannot find happiness. Those born so weak that they cannot survive, certainly cannot be said to have a created right to live. In nature it's all the can of ability to do. No person of any education and unbiased mind can deny the absolute truthfulness of that statement. Nature is one ceaseless struggle, equilibrium frequently resulting through active cataclysm. Thus far in the evolution of our race the social equilibrium has only resulted in the same way. The idea of liberty of our fathers assumed a *can* which is not possible under natural or existing social conditions. All people are not born equal by any means, so the natural result is that only those who have the might to can, can attain to anything approaching happiness. Happiness, what is it? We look into the dictionaries and they give us the very unphilosophical definition, "a condition of contentment." A definition which has not a self-evident reason is worse than none at all. Happiness is indeed a condition, but it is not "a condition of contentment." Such a condition, if of any possible degree of permanency, would soon lead to the demoralization of the race. Contentment almost presupposes inaction, whereas life is action. Happiness is something induced. It may be defined *as that mental condition of the individal in which the actor is so busied in some agreeable undertaking as to be unconscious of self-existence.* That definition is applicable to all sorts of conditions of happiness, regardless of cause. The moment consciousness of existence, without some definite action, occurs, the individual soon becomes unhappy. The majority of people are not in that fortunately equable social condition that they care to be conscious that they really live. The margin between life and living is too narrow. Were they all born "equal," all should be either equally happy or equally miserable. The equality and fraternity of men may be an evolutional phylogenetic fact; that is, as to the primary origin of the race, but there the simile ends. The inequality and want of fraternity makes itself apparent with the development of the can of might in life's struggle. It is because of the fact that not all who are born can live on account of the unequal distribution of the might to live that we have so much unhappiness.

Man is, however, when normally developed and among civilized people, a more or less highly intellectual result of forces which have been at work for untold generations. He is conscious of his surroundings and of his weakness and strength. Very few, however, have the psychical attribute of logical reasoning. A great many somehow believe that a "brotherhood of man" should exist, without ever thinking that such a condition is a rank impossibility, unless the time comes when all men are actually born equal and with such superhuman intelligence as to be able to completely nullify the "struggle for existence" by the artificial equilibrium of all conflicting social forces. We are all not only children of our time, but of times more or less immediately preceding it. As repeatedly said, the liberty and equal rights idea of the last century is still poignant in this. The veneration of the fathers has made it almost a sacred principle with the majority. It failed, and fails in not recognizing that justice is a greater principle. It failed in utterly neglecting to recognize the natural inequalities in man and the existence of that pregnant fact, the "struggle for existence." The idea of fraternity is the result of an entirely unintelligent and illogical attempt to ameliorate the painfulness of the struggle for those who have not the might to can. It is based on a slowly passing superstition, the result of ignorance. Darwin had not pointed out the hard and narrow road to truth in the days of our fathers. The natural struggle for existence, with its individualistic survival of the fittest, and heartless but slow perishing of those who have not the might to can, will and can only be equalized when humanity has evolved to that millennial condition of superfine intelligence, that sexual selection is so rigidly controlled that only those are produced who equally have the might and can equally survive in the conditions thus gradually developed. Not "fraternity," but that degree of individual excellence which renders all charity unnecessary is the tough road over which we must pass before all men are created equal and can with equal ability pursue happiness. Man alone can by the might of a sublime, profound, and mighty intelligence create this desired social condition. Socialism is a science. That is its mission.

Now, the result of the doctrine of freedom of equal rights has been to make all men at once assume them as their "inalienable rights," irrespective of their ability to enforce them from their environments. The fathers declared these rights to all men and unsuccessfully endeavored to make them practicable in the constitution. They were unsuccessful because their attempts were in direct contradiction to natural results. We cannot successfully contradict nature. They made America the land of promises neither they nor their children, nor their children's children, even to the ten-thousandth generation, can ever fulfill so long as sexual selection takes its unnatural course. The discontented of the world looked hopefully to America as the "promised land," for has not "Uncle Sam land enough to give us all a farm"? To be sure the land was here and is still here, but not all who ask for it have the natural might that they can make it their own and keep it. The unsatisfactory condition of the agricultural class is as natural as nature itself. As a class they have not the might to can in the same degree as those who have made themselves successful in the race for life and become that popular bugbear, "capitalist." Those who have the natural might can and generally do leave the farms for those pursuits which call forth all their might, because thereby they know they can live a more satisfactory existence. The farmers, in general, represent those weaker ones who eke out a miserable existence on the outskirts in the natural struggle for existence. Those among them who have the might to can are not found in the popular uprising. It is a hopeful sign for the future that the others have become aware of the fact that something is wrong with them. The same is true with the wage-earner. They both have taken the first natural step to ameliorate their condition. They recognize their individual weakness, and so are pooling their necessities. They are opposing the might of mass-power in hope that they *can* thus overcome the force of the might of natural individualism. It is a wise move, if tempered with just intelligence. The result in the end will come out all right. They have not, however, learned the first lesson. They must learn first that the weakness lies in themselves individually. The inequality is theirs. It is of nature. They were so created by their parents and their parents' parents. They must educate themselves not only to breed better, but only so many as they can so educate, that in both breeding and education they are equal to the severity of their environments. They can learn a valuable lesson by studying the families of those who have the might to can and succeed in the social struggle.

Let the poor farmer at home study the family of his rich brother in the city. Let the unsuccessful laborer study the same of the other fellow who "came over in the same ship" and who was then equally poor except in that one winning attribute, the might to can. Self-study is the first necessity in this great social problem. Self-study will solve the problem of the low social condition of the masses. Legislation, the heaven-hoped-for panacea of the masses, can do but little for them. The solution of the social problem is not to be sought in free silver, nor in free trade, nor land nationalization, as such, alone or united, *but in self-study, education, and in rendering the struggle for existence an impossibility* by breeding only those who have the might to can without severe struggle. Self-made men never existed. They are all born with the might to can, so that it is really no very bitter task for them to succeed. The struggle for existence is really a healthy stimulus. They are the "thoroughbreds" on the human race-course. The others are the "duffers." It is time they recognized this fact. It is time that they had intelligent responsibility enough to exercise sexual selection enough not to produce the same degree of "duffers" in the next generation. *This is the real key to the social problem. Intelligent sexual selection by man and woman, the fewest possible number of children, and then educate, educate, educate, in the same direction.*

The real sufferers in the unequal social struggle are the ones most heedless in their sexual relations and in the irrational production of their own kind. They forget that "likes beget likes," not only in form and species but characteristics also. The phenomenal "self-made" accidents in such breeding only more shockingly prove the rule. The more careful sexual selection and more intelligent limitation of children, with the self-evident advantages of better education also, is proving, that "like begets like," though not in a very marked degree in those who have the might to can.

To return again, and finally, to the effect of the "liberty and equal rights" idea of our fathers on our present civilization. No American who has not traveled largely and at the same time made a close study of the opinions of European people, high and low, can have any idea what conceptions they have of conditions in this country. The high-born (that is the fortunately born) generally have an idea that this is a sort of lawless country; while the others have the same idea, though in a different way. At home they think it is the laws alone which keep them back, as the same class largely do here, while they look on America as the "land of the free," where they can do as they please and then think they must succeed. They do not stop to think that American civilization is largely a product of the very institutions and conditions which they are unable to combat in Europe. They come to us and make the painful discovery that "things are worse here than in the old country," simply because, though approximately the same in fact, they are presented to them in changed or unaccustomed forms. They look for the ideal "freedom" in the "promised land," but find it not. Capitalistic employers are just as exacting in buying labor in the cheapest market; landlords demand all the rent they can get; store-keepers want their money as punctually, and wives have just as many children as they did in the old country. In no way is the promise fulfilled. Those who have the might to can succeed, as illustrated in hundreds of cases, while those who have not fall even lower in the social scale than they would had they remained at home in their more customary and and hence friendly environments. They find that "liberty, equality, and fraternity" are myths, notwithstanding the vast amount of charity and kindly feeling among us. They find that the fathers, whose names they may have revered as sort of superhuman men—and they were truly all they think they were—promised more than they could give, when they solemnly declared:

We hold these truths to be self-evident, that all men are created equal; that they are endowed by their Creator with certain inherent and inalienable rights; that among these are life, liberty, and the pursuit of happiness.

On the contrary, they soon learn that such rights do not exist, that rights can only come as forced concessions, the result of organized mass-might. Hence our organizations, trade unions, farmers' alliances. The lesson all must learn is, that equitable social conditions can only result from the most intelligent study of all these questions from the foundation of society up, from man's earliest beginning to the present; that justice and justice only is all that is needed; that there is no such thing as equal creation, but that all can and should take an equal part in intelligently balancing apparently opposing forces, all of which, however, have a common and closely interlacing interest; for each citizen in the nation is but a part of the whole machine, as well as does each interest bear most closely on every other interest in the prosperity of the general interest. Seek, then, justice intelligently. Try to realize that "liberty, equality, and fraternity" are impossible ideal conditions buried in the fond hopes of a past century. Let the "dead past bury its dead." Let us adjust ourselves to the actual conditions of a living present.

THE SPIRIT OF THE TWENTIETH CENTURY.

The time has come for a change. The rumbling of the wheels of justice can be distinctly heard in many lands. Thomas Paine found, to his bitter sorrow, that, after having done more for the cause of humanity than any man who has ever lived, after having suffered imprisonment for the sacred cause for ten weary months in a French prison, escaping death on the "altar of liberty" by the accidental closing of the door of his cell, that, on turning once again to that land where the "tree of liberty" had been so gloriously planted, the tree was barren of fruit, and even for the great son of liberty there was nothing left but neglect and persecution for his declining years. Thus it was with Garrison, Parker, Phillips, and Sumner, later prophets in the same field. Thus it is to-day, because it has not been understood that liberty is not consistent with justice. Liberty and individualism are one and the same thing. Their principle is of the "go-as-you-please kind," utterly regardless of the welfare of others. Liberty is the might of power untempered by the mitigating influence of nobler justice.

Theodore Parker is credited with being the author of the expression that a democracy is "a government of the people, for the people, by the people." Parker, intelligent and logical as he was, did not see that such a government as that could only end in disaster, robbing the people of their liberties, simply because that ideal can only be attained when a millennium of universal education and profound justice has been reached. The true idea of a democracy should read "a government of the people, for the people, by the selected intelligence among the people." Americans are very apt to over-estimate the true value of their government. This country has developed to its present phenomenal condition more in spite of the government than through any intelligent assistance it has received from it. We forget the wonderful natural resources open to the energies of man in almost every direction; a stimulus to the most intelligent and untiring endeavors, because of the fruitful reward of labor. We forget that, until perhaps the last thirty or forty years, the supposed liberty and these great natural resources offered by this country attracted to its shores the most active and energetic brains of Europe; the world conquerors, if I may use the term. We forget that up to that time the country was large, the population not over dense, and that to combat nature and win her rewards called forth all the inventive energy of the people, because it was largely every man for himself, and labor could not be had, so that machinery had to be invented to overcome the difficulties. We forget that while the conditions mentioned attracted the active and aggressive brain of Europe in this practical direction of looking out for self, that that still higher quality of the human mind which looks out for for others, the scientific in every sense, has found little or no attraction in the crudities and ceaseless struggle of American life. In this regard, in developing men so bred and educated that they could only give their lives to the study of all questions bearing directly on the welfare of man as a whole [science], we have been and still are woefully behind the civilization of the old world.

Only on questions related to human rights have we produced any marked statesman, and then only twice in our history. In 1776 the "rights of man," led off by Thomas Paine, became the slogan cry of the country. It called out and developed a large number of the most eminent leaders that the world has ever known. They did their utmost to establish the "rights of man" in the constitution, but, strangely to say, neglected entirely to consider that the black man had any "rights," which called forth again the great leaders of the antislavery movement, which culminated in the war of secession and the sanctification of the rights of all men to a use of their natural faculties untrammeled by legal interference, so long as the use of those faculties was just regarding other men, in the blood of the nation. Then came a standstill. Then a retrograde movement, which has been going down, down, into the depths of the most demoralizing slavery the world has yet seen—the willing subjection of a constitutionally free people to the chains of political slavery.

The pool of political corruption is stagnant enough on its surface, and foul with the stench of machine rottenness, but down deep the people are still pure, awaiting only wise, studious, and determined leaders to declare again the rights of man. May it be a peaceful declaration and not darkened by the sad and bloody fields of all previous attempts in the same direction! It will take profound wisdom and great patience to avoid bloodshed in the coming conflict. For coming it is, more wide and diffusely extended than the first uprising of humanity one hundred years since.

The mistake of American democracy since the first few decades of its establishment has been and still is that its representatives in the government have only represented the average knowledge of the people. I use the word "knowledge" purposely. There is a vast difference between the intelligence and actual knowledge of a people. It would be an insult to any people to assert that they were so unintelligent that they could not educate themselves to a higher and better knowledge of things. It is not an insult to say to them that they are not, as a people, sufficiently educated to govern themselves wisely and successfully. The anarchaic conditions existing to-day between the capitalists and the wage-earners, between trades unionism and free labor, between the agricultural classes and capitalists, all go to show that the election of men of the masses, representing the average knowledge of the masses, renders our government almost imbecilically impotent in its treatment of these questions.

Discouraging as it may appear, it is a fact that republican institutions are the only ones in the history of the world which do not recognize the highest and strongest intellectual development as a necessity to wise and safe government in all its departments. Only at the two times which threatened seriously our national existence has the real wisdom of the country been called into our legislative halls.

The appearance of the black cloud of slavery on our national horizon almost at the same moment with the final settlement of the revolution called at first a few men, then more and more, until it culminated in the war of secession. After the storm comes the calm of deceptive peace on the surface of the treacherous seas, whether the blue ocean or the deep sea of social discord. But be not deceived! The seething caldron of discordant forces still moves hither and thither beneath the apparently calm surface. It is all quiet at Homestead now, but what volcanic uneasiness there is hidden beneath the deceptive peace.

Wisdom of the most profound type, education of the most broad character, will be soon demanded to quell the coming storm and prevent a revolution, not only here but in all Europe, such as the world never saw. Safety lies in the selection of the wisest and most just men in the land, not to represent parties as at present, but men rising out of the varied and conflicting interests, thoroughly understanding the needs of those interests, and equally well comprehending the value and relation of each special interest to all other interests, to adjust the political compass and navigate the "ship of state" safely through the storms and between the shoals of danger in the great social struggle now opening on us. It is a singular phenomenon in the evolution of humanity to see the ship of state placed in the command of the semi-ignorance of the land rather than in that of its most profound wisdom.

Science is the study of nature. Nature constantly works according to fixed and unchangeable laws. There is such a thing as social science, which includes political. It embraces the study of man in all his relations to nature, which includes himself, not only as man of today, but as man representing but a factor in the evolution of all species. Have we a single statesman so educated in our legislative halls today? The best we have are lawyers, knowing little of other things, but full of that abomination of all abominations, a fanatical respect for human precedents. Our legislation is so superficial, so patch-work like, as to be a disgrace to profound knowledge.

We hear of a "Jeffersonian democracy," which in its mistaken advocacy of individual liberty and state individualism cursed this country with a civil war, the effects of which are one of the prime factors in hastening on the present conflict for the "rights of man." We are told to bow down in humble worship of "Jacksonian democracy," which again cursed us with the demoralizing idea that "to the victor belongs the spoils," that robs every public servant of the people of his constitutional rights to a free expression of his thoughts, and corresponding action, in things political. No man in the history of this country was ever more undemocratic than Jackson. No one man, not even Jefferson Davis, ever promulgated such an accursed doctrine as that which robs a man of his manhood and makes a mere slave of him because he holds a public trust. No man ever sent so demoralizing a principle out among the people as Jackson, which makes every dead-beat in the country, every truculent relation of a congressman, or one of his henchmen, a ravenous and unscrupulous wolf, caring nothing for the public welfare so long as he can place his carnivorous fangs on the public purse and become a useful spoke in that dangerous machine which grinds a free people, constitutionally, into a nation of abject slaves, moving only at the bid of the sharp and unscrupulous "boss of the ring."

What we need, and it is coming fast enough, is some great uprising of the people in the cause of justice. Great necessities invariably call forth the intellectual giants to meet them. The necessities of national existence, which require the most profound study of every interest of the people, have been more of a blessing in many ways to the nations of Europe than the deceptive peacefulness of American civilization. They have been the means of placing monarchial governments far ahead of republican in their fitness to meet the coming storm. They have necessitated calling the very best men in the nation into their governing bodies. With all his eccentricities it is doubtful if we have as wise and close a student of the great social questions facing us to-day among our public men as William III of Germany. We need this turmoil to make politics a science instead of a checker-board among us. We need it to force our higher educational institutions to do their duty and educate our youth to be fit citizens of a country destined to be the battlefield in this war of social science. No one need be alarmed as to the final result. Cataclysms may come, but out of the turmoil and strife bright justice will spring forth.

The very fountain of all social science is not taught as a science in any of our higher educational institutions, and I do not think in any in the world. I mean biology applied to social conditions.

No people can expect to make a success of self-government until they have learned where they came from, how the race has developed, and how they themselves have been produced. We are too conceited. Theology has thus far been the ruin of the human race; the greatest bar to progress. It will continue to be until it becomes an historical curiosity of the dark ages. It has made man think he was created an intelligent, developed, and moral being—a finished machine. It has not yet been discovered that we are not a finished machine. No engineer can run his machine unless he knows all about its structure and the relation of part to part. We assume that we are capable of self-government when not a man in a million knows anything of the machine he is intellectually supposed to control. A republican government has its foundation in the assumption that each individual knows enough to govern himself. The exception is so gigantically rare that the rule that no one knows enough to do this is almost a law. It will take bolder men, men of a higher type than our educational guardians dare at present obtain, men of nerve and intellectual courage in our highest colleges, before this desired end can find its beginning. They must be men who know, men above belief or creed. They must have the martyr spirit or they will be useless. The time for such education is now. It only can save us from the black dangers of social revolution which threaten us. The ordinary instructor is but a poor manipulator of musty error. He is too far back in the highest intellectual qualities to speak the truth without fear or favor. Those useless branches, except to specialists, Greek and Latin, should be dropped and the social sciences put in their place. Humanity will advance. Evolution is too plainly marked in every tendency of the times to be denied. The only cause for discouragement is the danger of revolution.

Only in republican nations do we see the fate of the people carelessly entrusted to the representative ignorance and selfishness among the people. When the people once learn that they must be represented by the most profound wisdom, instead of the common-place, among them, then the danger line will have been passed.

The various and varied invested interests of this country, as said before, must each produce its special men capable of intelligently representing each interest, but wise enough to know that every interest is in some way dependent on every other interest for its success. The same is true of the wage-earner. I have actually no sympathy with organized labor, when I see it try to kill out all the manhood in man by making him a mere machine and fixing the price of labor equal for dolt and genius. I detest it when it says to me, "You must be one of us or you must starve." No body of men has had, or ever will have, wisdom enough to formulate a creed or principle broad enough to meet the wants of a free and thinking man. On the other hand, the wage-earners should unite and intelligently discuss their own interests; they should unite as they now have, according to their respective trades; they should thus interestedly unite and declare themselves free from the kings of party machines, as our forefathers did of all earthly kings; they should break their bonds of slavery and become true democrats; they should discover men in their midst, knowing their wants, knowing their relations to other interests, including the vested interests of the land, and elect such men to represent them. The same is true of the agriculturists. A beginning has been made. Let the good work continue. Such men as Powderly, Gompers, Arthur, Henry George, and others should be in our congressional halls, and even the deluded anarchist should have representation. Then the sting of the serpent of danger will be removed. There is nothing like "freedom of speech" to let off the accumulated steam in the revolutionary social teapot.

This is what I meant when I said I was a "democrat from Democratville," to which the secretary of agriculture enters his objections. It is rather amusing that one of the great Tammany leaders should have once said to me, "It is such democrats as you are who are ruining the future of the democratic party in this country." I am glad the "boss" is so well aware that there are enough of us to be dangerous to the slave-making machine. As the poet said to Abraham Lincoln, also a "democrat from Democratville,"

> "We are coming, coming, coming,
> A hundred million more,"

for

> "From Greenland's icy mountains,
> From India's coral strand,
> The sons of earth are waking
> To declare the just rights of man."

Patho-Biological Laboratory,
State University of Nebraska,
Lincoln, August 1, 1892.

INDEX.

INDEX.

www.ingramcontent.com/pod-product-compliance
Lightning Source LLC
Chambersburg PA
CBHW021502210326
41599CB00012B/1101

9783337325350